SpringerBriefs in Applied Sciences and Technology

SpringerBriefs present concise summaries of cutting-edge research and practical applications across a wide spectrum of fields. Featuring compact volumes of 50 to 125 pages, the series covers a range of content from professional to academic.

Typical publications can be:

- A timely report of state-of-the art methods
- An introduction to or a manual for the application of mathematical or computer techniques
- A bridge between new research results, as published in journal articles
- A snapshot of a hot or emerging topic
- An in-depth case study
- A presentation of core concepts that students must understand in order to make independent contributions

SpringerBriefs are characterized by fast, global electronic dissemination, standard publishing contracts, standardized manuscript preparation and formatting guidelines, and expedited production schedules.

On the one hand, **SpringerBriefs in Applied Sciences and Technology** are devoted to the publication of fundamentals and applications within the different classical engineering disciplines as well as in interdisciplinary fields that recently emerged between these areas. On the other hand, as the boundary separating fundamental research and applied technology is more and more dissolving, this series is particularly open to trans-disciplinary topics between fundamental science and engineering.

Indexed by EI-Compendex, SCOPUS and Springerlink.

Tin-Chih Toly Chen

Explainable and Customizable Job Sequencing and Scheduling

Advancing Production Control
and Management with XAI

 Springer

Tin-Chih Toly Chen (ID)
Department of Industrial Engineering
and Management
National Yang Ming Chiao Tung University
Hsinchu, Taiwan

ISSN 2191-530X   ISSN 2191-5318 (electronic)
SpringerBriefs in Applied Sciences and Technology
ISBN 978-3-031-85373-9   ISBN 978-3-031-85374-6 (eBook)
https://doi.org/10.1007/978-3-031-85374-6

This Springer imprint is published by the registered company Springer Nature Switzerland AG
The registered company address is: Gewerbestrasse 11, 6330 Cham, Switzerland

If disposing of this product, please recycle the paper.

Competing Interests The author has no competing interests to declare that are relevant to the content of this manuscript.

Contents

Chapter 1
Explainable Artificial Intelligence (XAI)

1.1 Explainable Artificial Intelligence (XAI)

Artificial intelligence (AI) includes various technologies, such as machine learning (ML) and deep learning (DL) technologies that enable computers to imitate human behavior [1]. The computing speed, storage capacity, reliability, and interconnectivity of computers combined with human reasoning patterns give AI the ability to solve complex and large-scale problems. However, some AI applications, especially those based on DL technologies, are not easy to understand and communicate [2], making users unable to use the related applications with confidence. **Explainable artificial intelligence (XAI)** therefore emerged and has become a new trend in AI [3]. The recent emergence of generative artificial intelligence (GenAI) to communicate with users in natural language can also help solve similar needs [4].

According to Defense Advanced Research Projects Agency (DARPA) [5], XAI is to generate more explainable ML models, while maintaining a high level of learning performance (prediction or classification accuracy). XAI enables humans to understand, appropriately trust, and effectively manage these AI applications or their agents. General Data Protection Regulation (GDPR) that has been implemented in European countries defines the concept of XAI as customers having the right to interpret decisions made through automated systems [6]. However, what kinds of AI technologies are explainable is inconclusive. In the view of McNamara [7], ensemble methods, random forecasts, decision trees, Bayesian networks, sparse linear models, and others are highly explainable, while artificial neural networks (ANNs), DL, type-n fuzzy logic, support vector machines (SVMs) and others are less explainable. Therefore, XAI techniques and tools are referred to as white boxes or the twins of AI applications that are often considered black boxes [8].

So far, XAI development has emerged in at least three directions (see Fig. 1.1):

- XAI is to enhance the practicality of an existing AI technology by explaining its reasoning process and/or result [5]. For example, the concept of local interpretable

T. T. Chen, *Explainable and Customizable Job Sequencing and Scheduling*, SpringerBriefs in Applied Sciences and Technology, https://doi.org/10.1007/978-3-031-85374-6_1

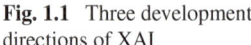 **Fig. 1.1** Three development directions of XAI

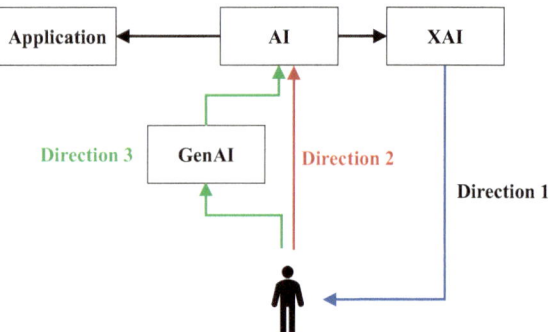

model-agnostic explanation (LIME) is to explain the classification or prediction result using a ML (or DL) model by identifying critical features/predictors and fitting simple and locally interpretable models [9]. In LIME, a synthetic data set is generated, because raw data sometimes contain exceptional cases and may not cover the entire sample space, which may complicate the explanation model. Simple decision rules or linear regression equations have been adopted to explain the reasoning mechanisms of deep neural networks (DNNs) that are usually nonlinear [10, 11]. Some studies have used heatmaps to emphasize the parts of an image that a DNN emphasizes when classifying or recognizing patterns [12, 13].

- XAI is to improve the effectiveness of existing AI technologies by incorporating easy-to-interpret visual features such as heatmaps [12, 13], decision (or regression) rules [14], decision trees (DTs) [11, 14], scatter plots [15], etc., to help diagnose its reasoning mechanism. An example is attention-based DNNs that incorporate DNNs with heatmaps [12, 13]. However, the linkage between the visual feature and the reasoning mechanism needs to be discovered.
- XAI is to enhance the convenience and effectiveness of AI applications by using natural language processing (NLP), DL, and accelerated computing, namely GenAI [4], to make user queries more friendly and collect and reorganize Web data faster to fulfill specific tasks.

1.2 Implementation Procedure of XAI

Chen [16] decomposed the implementation procedure of XAI into seven steps, as illustrated in Fig. 1.2. For an AI technology application, the reasoning process and/ or result should be explainable and interpretable. In addition, the result needs to be validated. After these are done, the AI technology application can be said to be explainable, which contributes to its trustworthiness, fairness, accountability, and transparency [2, 8, 9]. In addition, the explainability of AI technologies fosters improvement ideas based on which these AI technologies can be modified to enhance their effectiveness [11–14].

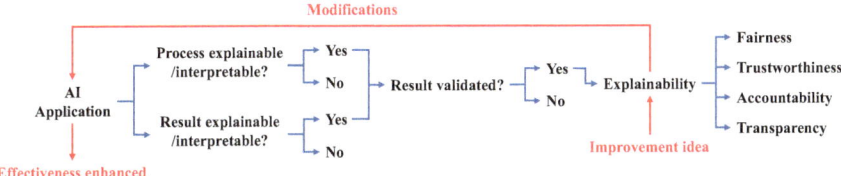

Fig. 1.2 Implementation procedure of XAI [16]

1.3 XAI Applications in Manufacturing

Some XAI applications in manufacturing are reviewed as follows.

Tirkel [17] built a feed-forward neural network (FNN) with two hidden layers to predict the cycle time of a job and compared the prediction accuracy with those achieved using an FNN with a single hidden layer and a DT. The FNN outperformed the DT, implying that the relationship between the cycle time and job attributes might be nonlinear. Moreover, by using multiple hidden layers, the prediction accuracy of the FNN was significantly improved. Confusion (or coincidence) matrices were also constructed to compare cycle time forecasts with actual values.

Chen and Wang [18] proposed a two-stage XAI approach to explain a classification-based cycle time prediction method. In their methodology, first, jobs are divided into several clusters. A scatter radar diagram is then designed to visualize the classification result. Compared with existing XAI techniques, the scatter radar diagram meets more requirements for better interpretation. Subsequently, an ANN is constructed for each cluster to predict the cycle times of jobs in the cluster. A random forest (RF) is then constructed to approximate the prediction mechanism of the ANN. In existing practice, a RF generates many decision rules to predict the cycle time of a job, which may cause confusion for the user. To solve this problem, they established a systematic procedure to re-organize these decision rules. In this way, the first few decision rules can provide most of the information, and the user does not have to read all the rules.

Shapley value (SHAP) analysis is obviously the most popular XAI technique for distinguishing the impacts of inputs on an AI application. For example, Kong et al. [19] constructed an ANN to predict the properties of an alloy based on its composition, for which SHAP analysis was conducted to compare the effects of components.

A similar treatment was taken in Akhlaghi et al. [20], in which a DNN was constructed to predict the performance of a dew point cooler according to its features. SHAP was applied to compare the effects of these features.

Attention mechanisms are another common XAI technique. Yeh et al. [21] constructed an encoder-decoder neural network to predict the quality measures of optoelectronic component products. To account for the correlations between quality measures, an additional decoder was added to the encoder-decoder neural network. Fang et al. [22] constructed a bidirectional long short-term memory (BiLSTM) neural network that can be applied to predict various time series data in a factory such as the unit costs, sales, yields, and cycle times of products. They designed a multimodal attention mechanism to improve the prediction performance by adding another hidden layer to the BiLSTM neural network to focus on more potentially useful historical data, which obviously made the BiLSTM neural network deeper and more difficult to understand.

Gao et al. [23] proposed an interpretable DL method called parsing key circuits through neuron synergy (PACER), in which the key neuron identification algorithm was used to comprehensively consider the global output contribution and local connection strength to select key neurons that most significantly affected the job output time estimation performance.

Senoner et al. [24] proposed an XAI method to improve the quality of a transistor chip, measured by the average and standard deviation of product yield. Various ML methods were used as AI applications to predict yields from many production parameter values, among which gradient boosting decision tree (GBDT) performed the best. Then, SHAP analysis was performed to determine the production parameters most critical to the transistor chip quality. Based on the SHAP analysis results, improvement plans for the most critical production parameters and the priority of executing these improvement plans could be formulated. Finally, they also conducted field experiments to verify the effectiveness of the improvement plans.

Wang et al. [25] constructed a two-dimensional LSTM model with multiple memory cells for predicting the job cycle time in a wafer fab. The LSTM was a three-layer recurrent neural network in which the outputs of hidden-layer nodes were fed back to themselves for subsequent predictions but might be forgotten. Hence, the LSTM was able to treat cycle times as a time series. The two dimensions in their model were used to account for the correlations between layers or wafers in duplicating LSTMs.

Chen and Wang [26] developed a fuzzy dynamic-prioritization agent-based system to improve the effectiveness of job cycle time prediction in a wafer fab based on the collaboration of multiple agents. Each agent built a fuzzy deep neural network (FDNN) with two hidden layers with different configurations. The threshold on the output node was fuzzified to ensure that the membership of an actual value in the corresponding fuzzy cycle time forecast was greater than the agent-specified threshold. Therefore, the spread of a fuzzy cycle time forecast naturally estimated the cycle time range. In addition, the authority levels of agents were determined by their past prediction performances and were therefore unequal. For this reason, the fuzzy weighted intersection (FWI) operator [27] was applied to aggregate the fuzzy cycle time forecasts by these agents. Traceable aggregation [15], an XAI for fuzzy group decision-making, was employed to illustrate the aggregation mechanism and results.

Wang et al. [28] built an FNN with two hidden layers to predict the cycle times of jobs. Following the concept of LIME [9], the FNN was approximated by a classification and regression tree (CART) [14] for global interpretation. For jobs classified into the same branch, fuzzy linear regression (FLR) [29] was used to estimate their cycle time ranges from job attributes to provide local explanations.

Chen et al. [30] proposed a fuzzy collaborative intelligence (FCI) approach to improve the precision of cycle time range estimation, in which a DNN was first built to accurately predict the cycle time of a job. A RF was then constructed to explain the DNN. Each decision tree of the RF was fuzzified to estimate the cycle time ranges of the jobs learned by the decision tree. A fuzzy collaboration mechanism was also established between decision trees to narrow the cycle time ranges. The novelty of their approach resided in that the RF was not applied to predict job cycle times but was used to explain and fuzzify the DNN without solving complex nonlinear programming (NP) problems.

Chen and Wang [31] proposed a fuzzy collaborative intelligence (FCI) approach to enable experts to choose flexibly from six evaluation models according to their points of view. FWI was then applied to reasonably aggregate experts' evaluation results. An XAI technique, traceable aggregation [15], was also applied to explain the aggregation process and result. The proposed FCI approach was applied to assist a precision machining factory evaluate and compare five innovative robotic applications in manufacturing. According to the experimental results, effectiveness was the most relevant criterion for innovative robotic applications. Furthermore, transferring machines from overcapacity to undercapacity factories under the guidance of an Internet of Machines (IoM) outperformed the other innovative robotic applications.

Figure 1.3 provides statistics on the fields in manufacturing where XAI techniques and tools are most commonly applied. Most commonly applied fields in manufacturing include quality control, decision-making, and maintenance. In contrast, job scheduling and inventory control are manufacturing fields where XAI is rarely applied. Therefore, the application of XAI techniques and tools is an urgent task and an opportunity for these fields.

Chen [16] divided the applications of XAI tools and techniques in manufacturing into the following categories: explaining the classification/estimation/optimization mechanism and result, explaining the composition and aggregation of criteria,

Fig. 1.3 Number of references about XAI applications in various manufacturing fields from 2010 to 2024 (*Data source* Google Scholar)

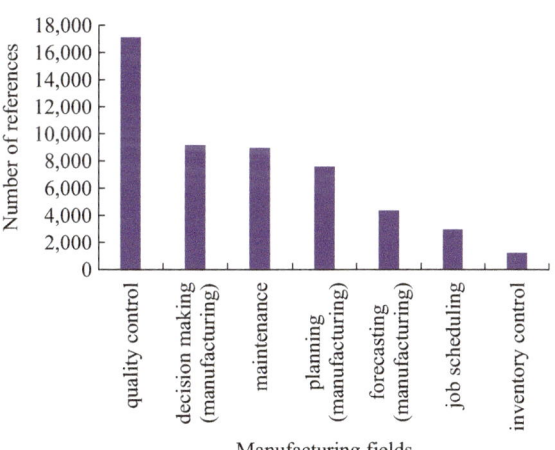

Fig. 1.4 Categorization of
XAI applications in
manufacturing

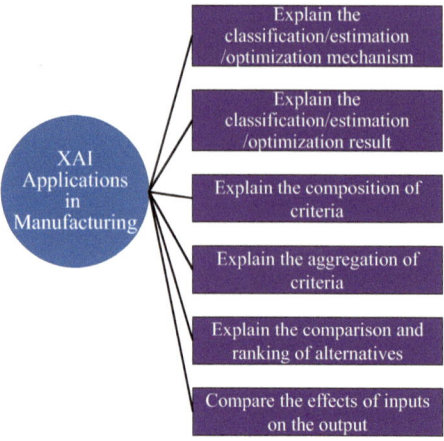

explaining the comparison and ranking of alternatives, comparing the effects of
inputs on the output, etc. (see Fig. 1.4).

1.4 Difficulties in Applying XAI in Manufacturing

So far, XAI has had many applications in manufacturing. However,

- The result of an AI technology application in manufacturing may be directly put
  into practice without human intervention, which eliminates the need to explain
  it to the related people. In contrast, the result of an AI technology application in
  healthcare or another domain often requires human approval or decision-making.
  Therefore, there are fewer XAI applications for explaining the reasoning process
  and/or results of AI applications (i.e., XAI applications in the first direction) in
  manufacturing, and most efforts have been placed on how to enhance the effectiveness of existing AI applications (i.e., XAI applications in the second direction),
  which may make the AI applications more like black boxes, while the application
  of GenAI in manufacturing is still in its infancy.
- In addition, engineers tend to apply AI technologies that are more sophisticated but
  potentially more effective. This phenomenon makes it more complex to explain
  or enhance AI applications in manufacturing.
- Furthermore, while there have been a few reviews outlining XAI applications that
  could be done in manufacturing [32–35], there aren't many real case studies.
- Most XAI applications in manufacturing are technology-driven rather than
  demand-driven. For example, applying Shapely additive explanation value
  (SHAP) analysis to select the most important production parameter, product
  component, machine status, and so on has dominated XAI research in the past

[19, 24], because R and Python provide relevant functions that can visualize the analysis results in an attractive way [36, 37].

- Past applications have only focused on limited functions in manufacturing, such as forecasting [19, 24, 28], quality engineering [19, 24, 38, 39], production optimization [40, 41], predictive maintenance [42], etc., while other topics, such as job scheduling [41] have not been well investigated.

1.5 Basic XAI Techniques

Some basic concepts in XAI are introduced as follows. A **local explanation** is to explain the rationale behind the result of a single example. A **global explanation** is to explain how the model arrives at its prediction/result and can be in the form of a visualization, a mathematical formula, or a diagram of the model. **Contrastive explanations** help us understand why a model generates a certain result/output for a given input and not another. Probably the most useful explanation technique is **what-if explanation**, which helps us understand the relationship between the model result and input features, similar to the concept of a parametric analysis. **Counterfactual explanations** tell us how changing assumptions about the inputs or parameters of a model can cause the model to arrive at a particular different result, similar to the concept of a sensitivity analysis. **Example-based explanations** are the simplest explanations in which the behavior of a model or underlying data distribution is simply explained by highlighting specific instances of the data. Methods for generating various explanations are summarized in Table 1.1.

A **model-agnostic XAI** method is that the computation required to explain an AI technology application is not based on the parameters of the AI technology application. **A post hoc interpretability method** explains the result, while **a pre hoc interpretability method** explains the process. Kamath and Liu [16] reviewed statistical and data analysis methods and tools suitable for the explanation of AI applications. In particular, they divided DL explanation methods into three categories: intrinsic methods, perturbation methods, and gradient/backpropagation methods.

Venugopal et al. [18] define the concept of **explainability failure** as a situation in which the output of an AI technology application is correct, but cannot be explained.

1.5.1 Lime

LIME is one of the most popular XAI techniques. LIME minimizes the following objective function:

$$Min \, \xi(x) = \arg \min_{g \in G} \mathcal{L}(f, g, \pi_x) + \Omega(g), \tag{1.1}$$

Table 1.1 Methods for generating various explanations

Type of explanations	Methods
Local explanations	• LIME • SHAP analysis • Etc

Global explanations	• System diagrams
	• Flowcharts
	• Mathematical formulae
	• Pseudocodes
	• Performance measures
	  – Forecasting: Mean absolute error (MAE), mean absolute percentage error (MAPE), root mean squared error (RMSE), R^2
	  – Classification: Recall, precision, F1, true positive rate (TPR), false positive rate (FPR), receiver operating characteristic (ROC) curve, area under curve (AUC), hit rate, cross-entropy loss
	  – Decision-making: Objective function value
	• Etc

Contrastive explanations (or counterfactual explanations)	• Contingency table analysis • Cross-tabulation • Local foil tree method • Etc
What-if explanations	• Decision rules • Sensitivity analysis • Parametric analysis • Etc
Example-based explanations	• CBR • kM • Fuzzy c-means • Etc

where

- f: the original model;
- g: the interpretation model;
- π_x: the proximity from all approximate outputs to the original output;
- $\arg \min_{g \in G} \mathcal{L}(f, g, \pi_x)$ returns the g value that minimizes $\mathcal{L}(f, g, \pi_x)$;
- $\Omega(g)$ is a measure of model complexity;

  LIME is implemented according to the following steps:

- Pre-model data analysis, using techniques such as bar-and-whisker plots, density plots, violin plots (for verifying data distribution), correlation analyses, correlation matrixes, heatmaps (for discovering data relationships), and backward regression elimination (for identifying most important inputs).
- AI model training.

Fig. 1.5 LR equations for fitting the synthetic data in local feasible regions

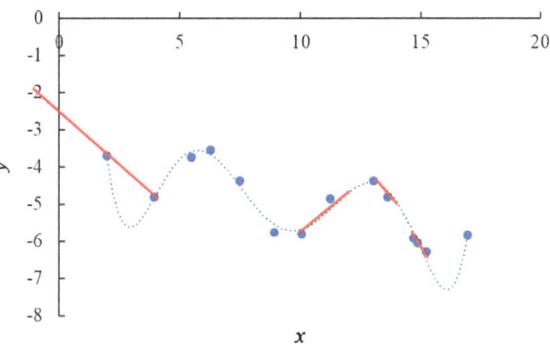

Fig. 1.6 Decision rules for approximating local relationships with stepwise values

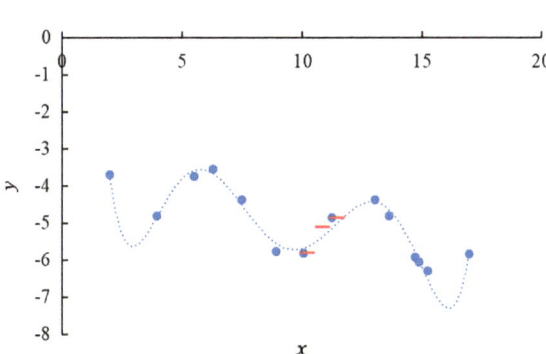

- Synthetic data generation: The most common way to generate synthetic data is randomization. For inputs with correlation, synthetic data can be generated by randomizing the major input before deriving the other. Professional expertise is important for generating synthetic data for special cases such as medical applications.
- AI application to synthetic data.
- Approximate AI application by fitting synthetic data: Although most previous studies have constructed linear regression (LR) equations to fit the synthetic data in local feasible regions (see Fig. 1.5), some recently constructed decision rules to approximate local relationships with stepwise values (see Fig. 1.6).
- Explanation performance evaluation.

1.5.2   SHAP Analysis

Post hoc XAI methods apply techniques such as partial derivation, odd ratio, out-of-bag (OOB) predictor importance, recursive feature elimination (RFE), permutation feature importance (PFI), SHAP analysis, etc. to evaluate the importance of each input to the output [16], thereby increasing the transparency and interpretability of the ML or DL model, of which SHAP analysis is probably the most popular.

The SHAP value is defined as the weighted average of the marginal contributions of all possible coalitions $|\mathbf{F}|!$ as [43]:

$$\varphi_m(f) = \sum_{\{S \subseteq F\} \setminus \{m\}} \frac{|S|!(|\mathbf{F}| - |S| - 1)!}{|\mathbf{F}|!} \cdot [f(x_{S \cup \{m\}}) - f(x_S)], \tag{1.2}$$

where

- $\varphi_m(f)$ is the weighted average Shapley value that feature m provides in the context of all coalitions that exclude feature m;
- $\mathbf{F}$ is the set of all features;
- S is a subset (i.e., coalition) of $\mathbf{F}$;
- $f(x_{S \cup \{m\}})$ is the model prediction considering feature m, while $f(x_S)$ is the model prediction without considering feature m.

A special case is when $|S| = |\mathbf{F}| - 1$, i.e., $S = \mathbf{F} \setminus \{m\}$,

$$\varphi_m(f) = \sum_{\{S \subseteq F\} \setminus \{m\}} \frac{|S|!(|\mathbf{F}| - |S| - 1)!}{|\mathbf{F}|!} \cdot [f(x_{S \cup \{m\}}) - f(x_S)]$$

$$= \sum_{\{S \subseteq F\} \setminus \{m\}} \frac{|S|!0!}{|\mathbf{F}|!} \cdot [f(x_{S \cup \{m\}}) - f(x_S)]$$

$$= \frac{\sum_{\{S \subseteq F\} \setminus \{m\}} [f(x_{S \cup \{m\}}) - f(x_S)]}{|\mathbf{F}|} \tag{1.3}$$

In this way, SHAP evaluates the importance of an input by fixing its value while randomizing other inputs [44], and then averaging the resulting differences in $f()$. Inputs to a scheduling problem include processing times, machine capacities, release times, due dates, job sizes, job priorities, operation sequences, etc. Most of these cannot be meaningfully randomized. The results of a SHAP analysis are often summarized with a tornado plot (see Fig. 1.7). As can be seen from the figure, the absolute value of the SHAP value of x_2 is the highest, so x_2 is the most important input, followed by x_4. However, the SHAP value of x_2 is positive, which means that by setting x_2 to the current value, the output has increased relative to the average.

1.5.3 Local Foil Tree Method

The local foil tree method constructs a decision tree centered on any data point in question. Decision trees are trained to locally distinguish foil classes from any other

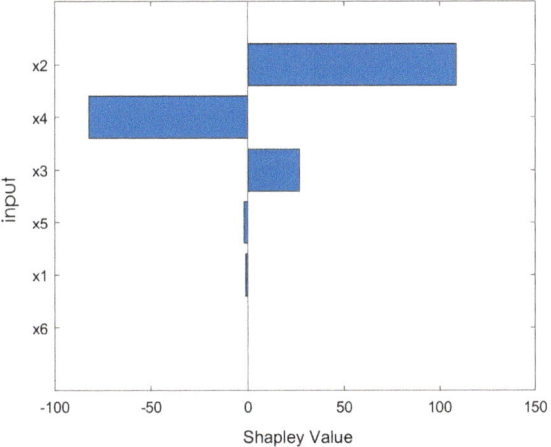

Fig. 1.7 Tornado plot for summarizing the results of a SHAP analysis

class, including fact classes [45]. The local foil tree method consists of the following steps [45]:

- retrieve the fact;
- identify the foil;
- generate or sample a local dataset;
- train a decision tree;
- locate the "fact-leaf";
- locate a "foil-leaf";
- compute differences;
- construct explanation.

An example of contrastive explanations generated using the local foil tree method is given in Fig. 1.8.

1.5.4 Attention Mechanism

A prevalent XAI technique aiming to enhance the effectiveness of an AI application is to add an attention mechanism [46] to the AI model, which has shown its potential in various pattern recognition [47–50] and time series prediction problems [51–59].

In a recurrent neural network (RNN) model, the outputs from the output-layer nodes of the current example ($\mathbf{h}_t$) and the outputs from the hidden-layer nodes of every past example ($\mathbf{h}_s$) are used as the hidden states of the current example. Similarly, for an autoencoder, hidden states include outputs from the encoder and decoder nodes of the related examples. The hidden states are fed into a score function using scaled dot-product attention:

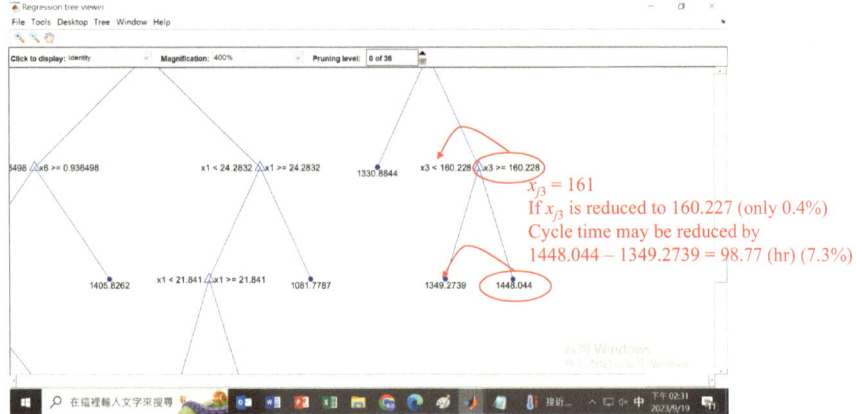

Fig. 1.8 Example of contrastive explanations generated using the local foil tree method

$$s(\mathbf{h_t}, \mathbf{h_s}) = \frac{\sum_{j=1}^{K} h_{tj} h_{sj}}{\sqrt{K}}.$$ (1.4)

The softmax function (σ) is usually used to evaluate the importance of each $\mathbf{h_s}$ to $\mathbf{h_t}$:

$$\sigma(s(\mathbf{h_t}, \mathbf{h_s})) = \frac{e^{s(\mathbf{h_t}, \mathbf{h_s})}}{\sum_{q=1}^{t-1} e^{s(\mathbf{h_t}, \mathbf{h_q})}}.$$ (1.5)

The results are between 0 and 1 and are weighted averaged with the original hidden states:

$$\mathbf{c_t} = \sum_{s=1}^{t} \sigma(s(\mathbf{h_t}, \mathbf{h_s}))\mathbf{h_s}$$ (1.6)

which becomes the new inputs to the output layer or are fed into a LR equation to derive the final outcome:

$$o_t = \sum_{l=1}^{L} w_l c_{tl}.$$ (1.7)

Fan et al. [51] constructed a BiLSTM neural network to forecast the sales of products for an online retailer. They designed a multi-modal attention mechanism to improve the prediction performance between the encoder and decoder of the BiLSTM neural network. During the decoding stage, the BiLSTM hidden state for each future time step attended to a different part of the history, thus forming a hidden representation of the future step. The temporal attention mechanism was applied to M periods of historical data and fused them with multi-modal attention weights obtained from the interaction of previous hidden states with transformed content vectors.

Shih et al. [52] constructed an LSTM neural network with an attention mechanism to forecast several time series, including solar power generation, road occupancy, electricity consumption, and exchange rates. In the attention mechanism, hidden states were extracted from the LSTM neural network and fed into a convolution neural network (CNN) to apply filters to the hidden states. The outputs of the CNN then became additional inputs to the LSTM.

Adil et al. [54] also constructed several BiLSTM neural networks to predict the arrival of tourists. Before prediction, the Loess (STL) method was applied to decompose tourist arrival data (a time series) into trend part, seasonal part, and residual part. The trend and residual parts were predicted using separate BiLSTM neural networks. Adil et al. [54] added two attention layers to each BiLSTM neural network: one for features and another for the time step dimension. After inputting features, the attention mechanism was applied before the BiLSTM layer, which assigned more weight to those features (or periods) that were more important in tourist arrival prediction and less weight to those features (or periods) that were less important. Although the attention mechanism used by Adil et al. [54] is easy to replicate, it changes the optimality of the original BiLSTM neural network, which requires retraining the BiLSTM neural network. In addition, the attention mechanism may input only the most important attributes into the DNN and discard other attributes, such as the treatment taken in Senoner et al. [24] and other past studies. A problem of Adil et al. [54] is that there are many ways to assess the importance of an attribute. It is unclear which evaluation method is most suitable for a specific application. Additionally, the most appropriate assessment method may differ for different examples.

Tian et al. [55] constructed a gated recurrent unit (GRU) neural network to predict wind power generation, which was also a time series problem. A dual self-attention module was also embedded in the gated recurrent neural network (RNN), in which the importance of inputs and hidden states were evaluated separately based on the bias terms of the hidden features. The original inputs and hidden states were then modified before being fed into the gated recurrent neural network to improve the prediction accuracy.

To forecast the sales of a new product, Ekambaram et al. [53] applied K-nearest neighbor (KNN) and an RNN and compared the prediction performances of the two methods. Since the image of a product was considered an influential factor in predicting its sales, an encoder-decoder model was also built to capture the compact embedding of the image and merge it with temporal features (i.e., past sales and exogenous features) before using the RNN as a decoder to make predictions. In addition, a self-attention mechanism was established to determine the relative importance of each temporal feature (i.e., temporal feature weight) for each time period based on the temporal features and the hidden state of the RNN, so that more important attributes should receive more attention.

Table 1.2 summarizes the features of some studies using attention mechanisms.

Table 1.2 Features of some studies using attention mechanisms

Method	Prediction problem type	Prediction method	Attention/ correction mechanism type	Attention/ correction mechanism parameters	Part of historical data attended
Fan et al. [51]	Time series	BiLSTM	Global, soft, model-specific	Hidden states, inputs	Multiple parts
Shih et al. [52]	Time series	LSTM, CNN	Global, soft, model-specific	Hidden states, inputs	Single part
Adil et al. [54]	Time series	BiLSTM	Local, soft, model-specific	Hidden states, inputs	Single part
Tian et al. [55]	Time series	GRU neural network	Global, soft, model-specific	Hidden states, inputs	Single part
Ekambaram et al. [53]	Time series	RNN	Global, soft, model-specific	Hidden states, inputs	Single part

1.5.5 RF-Based Incremental Interpretation (RFII)

In traditional LIME applications, LR equations are usually used to fit the DL models in local feasible regions. Some recent studies used decision rules instead. A RF can also be constructed for the same purpose; i.e., multiple regression trees simultaneously apply to the same local feasible region. In this way, a better approximation accuracy may be achieved, as illustrated in Fig. 1.9.

In a RF constructed to approximate a local feasible region, there are a number of trees. From each tree, a regression rule can be applied to predict the cycle time of a job. As a result, multiple rules are simultaneously applicable, which may cause confusion for the user. To address this issue, Chen and Wang [18] proposed RF-based incremental interpretation (RFII) to re-organize the regression rules derived using RF as follows:

Step 1.   Average the prediction results by regression rules as $\hat{o}_j$.
Step 2.   Set $q = 1$.

Fig. 1.9 LIME with RF approximators

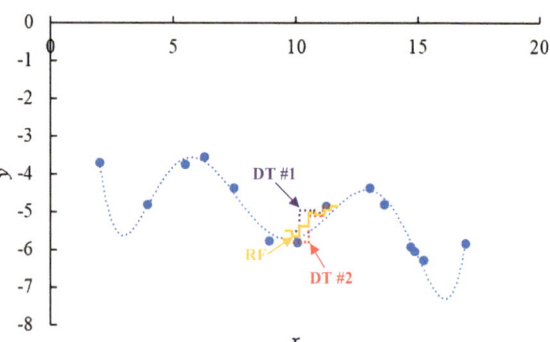

Step 3.   Choose the regression rule with the closest prediction result to $\hat{o}_j$.
Step 4.   Set the q-th incremental rule to the regression rule.
Step 5.   Set $\hat{o}_j$(current) to the prediction result of the regression rule.
Step 6.   Remove the regression rule.
Step 7.   Increase q by 1.
Step 8.   If $q > T$, go to Step 14; otherwise, go to Step 9.
Step 9.   Find the regression rule with the closest prediction result to $\hat{o}_j$.
Step 10. Subtract $\hat{o}_j$(current) from the prediction result of the regression rule, and divide the result by T: The result is indicated by $\Delta\hat{o}_j$(current).
Step 11. Set the premise of the qth incremental rule to that of the regression rule.
Step 12. Set the consequence of the qth incremental rule to "add $\Delta\hat{o}_j$(current) to $\hat{o}_j$" or "subtract $-\Delta\hat{o}_j$(current) from $\hat{o}_j$."
Step 13. Return to Step 6.
Step 14. End.

1.6 Organization of This Book

This book is intended to provide technical details on the development and application of XAI to job sequencing and scheduling, including methodologies, tools, system architectures, software and hardware, examples and applications. XAI is currently the hottest topic, since the acceptability of overly complex AI technologies is usually questioned. Popular techniques in job sequencing and scheduling include many bionic algorithms such as genetic algorithm (GA), artificial bee colony (ABC), and ant colony optimization (ACO). Most of their applications have not yet been widely explained, which undoubtedly limits the practicability of these methods. Nevertheless, simple XAI techniques and tools for explaining these AI applications are pursued. However, simple XAI techniques and tools often struggle to adequately explain an AI application. In other words, applying simple XAI techniques and tools to explain a complicated AI application in job sequencing and scheduling is a challenge. However, related research results are mostly scattered in various journal issues or conference proceedings, and there is an urgent need for a systematic integration of these results. To address these issues, this book systematically reviews the progress of XAI and introduces the methods, tools, and applications of XAI technologies in job sequencing and scheduling. Relevant references or real cases are also used as supporting evidence.

In specific, the outline of the present book is structured as follows.

The current chapter introduces the basic concepts of XAI, with an emphasis on manufacturing applications. First, the definition of XAI is given. The three current development directions of XAI are also mentioned. Subsequently, the implementation process of XAI application is established, including seven steps. Then some past research and applications of XAI in manufacturing are reviewed, thereby concluding that the most commonly used fields in manufacturing include quality control, decision-making, and maintenance, with the purpose of explaining classifi-cation/estimation/optimization mechanisms and results, explaining the composition and aggregation of criteria, explaining the comparison and ranking of alternatives, and comparing the effects of inputs on the output. Difficulties in applying XAI in manufacturing are also discussed. In addition, some basic XAI techniques are also introduced, including LIME, SHAP analysis, local foil tree method, attention mechanisms, and RFII.

Chapter 2, Artificial Intelligence (AI) Applications in Job Sequencing and Scheduling, reviews the applications of AI in job sequencing and scheduling. First, existing AI technologies are briefly categorized. Although symbolic AI has a wider application in job scheduling, data-driven AI is also suitable in this field. Subse-quently, some advanced AI technologies, such as AI 2.0 and generative AI, are also considered to have the potential to address job scheduling needs. Therefore, some applications of AI in job sequencing and scheduling are reviewed. A procedure is also established to apply AI technologies to job sequencing and scheduling. AI technolo-gies can be applied to facilitate several steps of job scheduling, and the performances achieved by these AI technologies often far exceed human capabilities. Afterward, the applications of some AI technologies in job sequencing and scheduling are given, including ANN, simulation, digital twin, GA, Industry 4.0, fuzzy logic, ACO, etc. Based on these examples, the problems faced when applying AI technologies to job sequencing and scheduling are discussed. Finally, the possible application of generative AI in job sequencing and scheduling is put forth.

Chapter 3, XAI Applications in Job Sequencing and Scheduling, introduces the applications of XAI techniques and tools in job sequencing and scheduling. First, such XAI applications correspond to various steps of the job scheduling process. Therefore, XAI applications for collecting and estimating the data required for job sequencing and scheduling are mentioned first. Applicable XAI techniques and tools include visualization XAI techniques and tools, XAI techniques for evaluating the importance of each input to the output, and XAI techniques for approximating the estimation mechanism. Subsequently, an introduction to XAI applications used to explain the scheduling mechanism is given. For this purpose, generic XAI tech-niques and tools, SHAP analysis, decision tree-based methods, LIME, and explain-able optimization are suitable. Finally, criteria for evaluating the effectiveness of XAI applications in job sequencing and scheduling are provided.

Chapter 4, Explaining Genetic Algorithm Applications in Job Sequencing and Scheduling, first explains the motivation for the application of GA in job sequencing and scheduling. Then the evolution process of GA is introduced, especially the steps. Several common techniques and tools for explaining or visualizing these steps are also reviewed. Clearly, such GA applications may remain difficult to understand and communicate due to their stochastic nature. XAI can play a role in solving this difficulty. The problems faced by existing XAI techniques in interpreting GA applications in job sequencing and scheduling are also discussed. To address these issues, XAI new techniques and tools have emerged. These techniques and tools fall into two categories: textual descriptions and visualization. Four visualization techniques for explaining the application of GA in job sequencing and scheduling are introduced, including generic visualization techniques, saliency diagrams, decision tree-based interpretation, dynamic transition and contribution diagrams, and Shapley value (SHAP) analysis.

Chapter 5, Explaining Other Bio-inspired Algorithm Applications in Job Sequencing and Scheduling, first highlights the fact that many bio-inspired algorithms, such as ANN, GA, ABC, ACO, PSO, and others, have shown their effectiveness in job sequencing and scheduling. However, these bio-inspired algorithms have long been considered black boxes, hindering the credibility and reliability of their applications in job sequencing and scheduling. Therefore, this chapter takes ACO applications in sequencing and scheduling as an example. Such applications have not yet been explained using XAI techniques, and recent successes in explaining the application of GAs in job scheduling can be replicated. First, past research on the application of ACOs in job sequencing and scheduling is reviewed. Motivations for applying XAI techniques and tools are then highlighted. Subsequently, emerging XAI techniques and tools for explaining ACO applications in job sequencing and scheduling are introduced, including a novel decision tree-based XAI approach. The examples reveal that most ants end up following similar paths. In addition, the contrastive gradient salient map precisely locates the convergence point of the ACO evolution process and intuitively displays it to the scheduler. Furthermore, the decision tree-based XAI approach fills the gap in explaining the intrinsic mechanisms of black-box job scheduling methods such as ACO.

References

1. T.C.T. Chen, Y.C. Wang, AI applications to shop floor management in lean manufacturing, in *Artificial Intelligence and Lean Manufacturing* (2022), pp. 75–90
2. S. Chakraborty, R. Tomsett, R. Raghavendra, D. Harborne, M. Alzantot, F. Cerutti, M. Srivastava, A. Preece, S. Julier, R.M. Rao, P. Gurram, Interpretability of deep learning models: a survey of results, in *IEEE Smartworld, Ubiquitous Intelligence & Computing, Advanced & Trusted Computed, Scalable Computing & Communications, Cloud & Big Data Computing, Internet of People and Smart City Innovation* (2017), pp. 1–6
3. D. Gunning, M. Stefik, J. Choi, T. Miller, S. Stumpf, G.Z. Yang, XAI—Explainable artificial intelligence. Sci. Robot. **4**(37), eaay7120 (2019)
4. M. Treppner, H. Binder, M. Hess, Interpretable generative deep learning: an illustration with single cell gene expression data. Hum. Genet. **141**(9), 1481–1498 (2022)
5. D. Gunning, D. Aha, DARPA's explainable artificial intelligence (XAI) program. AI Mag. **40**(2), 44–58 (2019)
6. W. Saeed, C. Omlin, Explainable AI (XAI): a systematic meta-survey of current challenges and future opportunities. Knowl.-Based Syst. **263**, 110273 (2023)
7. M. McNamara, Explainable AI: what is it? How does it work? And what role does data play? (2022). https://www.netapp.com/blog/explainable-ai/
8. C. Panigutti, A. Perotti, D. Pedreschi, Doctor XAI: an ontology-based approach to black-box sequential data classification explanations, in *Proceedings of the 2020 Conference on Fairness, Accountability, and Transparency* (2020), pp. 629–639
9. M. T. Ribeiro, S. Singh, C. Guestrin, Why should I trust you? Explaining the predictions of any classifier, in *Proceedings of the 22nd ACM SIGKDD International Conference on Knowledge Discovery and Data Mining* (2016), pp. 1135–1144

10. D. Kumar, A. Wong, G.W. Taylor, Explaining the unexplained: a class-enhanced attentive response (clear) approach to understanding deep neural networks, in *Proceedings of the IEEE Conference on Computer Vision and Pattern Recognition Workshops* (2017), pp. 36–44

11. T. Chen, H.-C. Wu, M.-C. Chiu, A deep neural network with modified random forest incremental interpretation approach for diagnosing diabetes in smart healthcare. Appl. Soft Comput. **152**, 111183 (2024)

12. E. Tjoa, H.J. Khok, T. Chouhan, G. Cuntai, Improving deep neural network classification confidence using heatmap-based eXplainable AI (2022). https://doi.org/10.48550/arXiv.2201. 000092022

13. J. Lee, H. Cho, Y.J. Pyun, S.J. Kang, H. Nam, Heatmap assisted accuracy score evaluation method for machine-centric explainable deep neural networks. IEEE Access **10**, 64832–64849 (2022)

14. H.-C. Wu, T. Chen, CART–BPN approach for estimating cycle time in wafer fabrication. J. Ambient. Intell. Humaniz. Comput. **6**, 57–67 (2015)

15. Y.-C. Lin, T. Chen, Type-II fuzzy approach with explainable artificial intelligence for nature-based leisure travel destination selection amid the COVID-19 pandemic. Digital Health **8**, 20552076221106320 (2022)

16. T.-C. T. Chen, Explainable artificial intelligence (XAI) in manufacturing, in *Explainable Artificial Intelligence (XAI) in Manufacturing: Methodology, Tools, and Applications* (2023), pp. 1–11

17. I. Tirkel, Cycle time prediction in wafer fabrication line by applying data mining methods, in *2011 IEEE/SEMI Advanced Semiconductor Manufacturing Conference* (2011), pp. 1–5

18. T. Chen, Y.C. Wang, A two-stage explainable artificial intelligence approach for classification-based job cycle time prediction. Int. J. Adv. Manuf. Technol. **123**(5), 2031–2042 (2022)

19. B.O. Kong, M.S. Kim, B.H. Kim, J.H. Lee, Prediction of creep life using an explainable artificial intelligence technique and alloy design based on the genetic algorithm in creep-strength-enhanced ferritic 9% Cr steel. Metals Mater. Int. 1–12 (2023)

20. Y.G. Akhlaghi, K. Aslansefat, X. Zhao, S. Sadati, A. Badiei, X. Xiao, S. Shittu, Y. Fan, X. Ma, Hourly performance forecast of a dew point cooler using explainable Artificial Intelligence and evolutionary optimisations by 2050. Appl. Energy **281**, 116062 (2021)

21. C.H. Yeh, Y.C. Fan, W.C. Peng, Interpretable multi-task learning for product quality prediction with attention mechanism, in *IEEE 35th International Conference on Data Engineering* (2019), pp. 1910–1921

22. W. Fang, Y. Guo, W. Liao, K. Ramani, S. Huang, Big data driven jobs remaining time prediction in discrete manufacturing system: a deep learning-based approach. Int. J. Prod. Res. **58**(9), 2751–2766 (2020)

23. P. Gao, J. Wang, R. Zhong, J. Zhang, Neuron synergy based explainable neural network for manufacturing cycle time forecasting. J. Manuf. Syst. **71**, 695–706 (2023)

24. J. Senoner, T. Netland, S. Feuerriegel, Using explainable artificial intelligence to improve process quality: evidence from semiconductor manufacturing. Manage. Sci. **68**(8), 5704–5723 (2022)

25. J. Wang, J. Zhang, X. Wang, Bilateral LSTM: a two-dimensional long short-term memory model with multiply memory units for short-term cycle time forecasting in re-entrant manufacturing systems. IEEE Trans. Industr. Inf. **14**(2), 748–758 (2017)

26. T.C.T. Chen, Y.C. Wang, Fuzzy dynamic-prioritization agent-based system for forecasting job cycle time in a wafer fabrication plant. Complex Intell. Syst. **7**, 2141–2154 (2021)

27. T.C.T. Chen, Y.C. Wang, C.W. Lin, A fuzzy collaborative forecasting approach considering experts' unequal levels of authority. Appl. Soft Comput. **94**, 106455 (2020)

28. Y.-C. Wang, T.-C.T. Chen, M.-C. Chiu, An explainable deep-learning approach for job cycle time prediction. Decision Analytics **6**, 100153 (2023)

29. T. Chen, The FLR–PCFI–RBF approach for accurate and precise WIP level forecasting. Int. J. Adv. Manuf. Technol. **63**, 1217–1226 (2012)

30. T.C.T. Chen, C.W. Lin, Y.C. Lin, A fuzzy collaborative forecasting approach based on XAI applications for cycle time range estimation. Appl. Soft Comput. **151**, 111122 (2024)

31. T.C.T. Chen, Y.C. Wang, Evaluating innovative future robotic applications in manufacturing using a fuzzy collaborative intelligence approach. Int. J. Adv. Manuf. Technol. **130**(11), 6027–6041 (2024)
32. G. Sofianidis, J.M. Rožanec, D. Mladenic, D. Kyriazis, A review of explainable artificial intelligence in manufacturing, in *Trusted Artificial Intelligence in Manufacturing* (2021), pp. 93–113
33. A. Kotriwala, B. Klöpper, M. Dix, G. Gopalakrishnan, D. Ziobro, A. Potschka, XAI for operations in the process industry-applications, theses, and research directions, in *AAAI Spring Symposium: Combining Machine Learning with Knowledge Engineering* (2021), pp. 1–12
34. I. Ahmed, G. Jeon, F. Piccialli, From artificial intelligence to explainable artificial intelligence in industry 4.0: a survey on what, how, and where. IEEE Trans. Ind. Inform. **18**(8), 5031–5042 (2022)
35. B.M. Colosimo, F. Centofanti, Model interpretability, explainability and trust for manufacturing 4.0, in *Interpretability for Industry 4.0: Statistical and Machine Learning Approaches* (2022), pp. 21–36
36. S. Lundberg, Welcome to the SHAP documentation (2018). https://shap.readthedocs.io/en/latest/index.html
37. P. Cacas, A gentle introduction to SHAP values in R (2019). https://blog.datascienceheroes.com/how-to-interpret-shap-values-in-r/
38. S. Meister, M. Wermes, J. Stüve, R.M. Groves, Investigations on explainable artificial intelligence methods for the deep learning classification of fibre layup defect in the automated composite manufacturing. Compos. B Eng. **224**, 109160 (2021)
39. A. Hanchate, S.T. Bukkapatnam, K.H. Lee, A. Srivastava, S. Kumara, Explainable AI (XAI)-driven vibration sensing scheme for surface quality monitoring in a smart surface grinding process. J. Manuf. Process. **99**, 184–194 (2023)
40. S. Perez-Castanos, A. Prieto-Roig, D. Monzo, J. Colomer-Barbera, Holistic production overview: using XAI for production optimization, in *Artificial Intelligence in Manufacturing: Enabling Intelligent, Flexible and Cost-Effective Production Through AI* (2023), pp. 423–436
41. Y.C. Wang, T. Chen, Adapted techniques of explainable artificial intelligence for explaining genetic algorithms on the example of job scheduling. Expert. Syst. Appl. **237**(A), 121369 (2024)
42. B. Hrnjica, S. Softic, Explainable AI in manufacturing: a predictive maintenance case study, in *IFIP International Conference on Advances in Production Management Systems* (2020), pp. 66–73
43. L.P. Joseph, E.A. Joseph, R. Prasad, Explainable diabetes classification using hybrid Bayesian-optimized TabNet architecture. Comput. Biol. Med. **151**, 106178 (2022)
44. Y. Meng, N. Yang, Z. Qian, G. Zhang, What makes an online review more helpful: an interpretation framework using XGBoost and SHAP values. J. Theor. Appl. Electron. Commer. Res. **16**(3), 466–490 (2020)
45. J. van der Waa, M. Robeer, van J. Diggelen, M. Brinkhuis, M. Neerincx, Contrastive explanations with local foil trees. arXiv preprint arXiv:1806.07470 (2018)
46. E. Hasanpour Zaryabi, L. Moradi, B. Kalantar, N. Ueda, A.A. Halin, Unboxing the black box of attention mechanisms in remote sensing big data using XAI. Remote. Sens. 14(24) 6254 (2022)
47. H. Jiang, Y. Lu, D. Zhang, Y. Shi, J. Wang, Deep learning-based fusion networks with high-order attention mechanism for 3D object detection in autonomous driving scenarios. Appl. Soft Comput. 111253 (2024)
48. D. Gao, W. Yang, P. Li, S. Liu, T. Liu, M. Wang, Y. Zhang, A multiscale feature fusion network based on attention mechanism for motor imagery EEG decoding. Appl. Soft Comput. **151**, 111129 (2024)
49. Z.B. Yang, J.P. Zhang, Z.B. Zhao, Z. Zhai, X.F. Chen, Interpreting network knowledge with attention mechanism for bearing fault diagnosis. Appl. Soft Comput. **97**, 106829 (2020)
50. P. Wu, X. Li, C. Ling, S. Ding, S. Shen, Sentiment classification using attention mechanism and bidirectional long short-term memory network. Appl. Soft Comput. **112**, 107792 (2021)

51. C. Fan, Y. Zhang, Y. Pan, X. Li, C. Zhang, R. Yuan, H. Huang, Multi-horizon time series forecasting with temporal attention learning, in *Proceedings of the 25th ACM SIGKDD International Conference on Knowledge Discovery & Data Mining* (2019), pp. 2527–2535

52. S.Y. Shih, F.K. Sun, H.Y. Lee, Temporal pattern attention for multivariate time series forecasting. Mach. Learn. **108**, 1421–1441 (2019)

53. V. Ekambaram, K. Manglik, S. Mukherjee, S.S.K. Sajja, S. Dwivedi, V. Raykar, Attention based multi-modal new product sales time-series forecasting, in *Proceedings of the 26th ACM SIGKDD International Conference on Knowledge Discovery & Data Mining* (2020), pp. 3110–3118

54. M. Adil, J.Z. Wu, R.K. Chakrabortty, A. Alahmadi, M.F. Ansari, M.J. Ryan, Attention-based STL-BiLSTM network to forecast tourist arrival. Processes **9**(10), 1759 (2021)

55. C. Tian, T. Niu, W. Wei, Developing a wind power forecasting system based on deep learning with attention mechanism. Energy **257**, 124750 (2022)

56. C. Yu, G. Yan, C. Yu, X. Mi, Attention mechanism is useful in spatio-temporal wind speed prediction: evidence from China. Appl. Soft Comput. **148**, 110864 (2023)

57. X. Liu, J. Zhou, Short-term wind power forecasting based on multivariate/multi-step LSTM with temporal feature attention mechanism. Appl. Soft Comput. **150**, 111050 (2024)

58. L. Su, L. Xiong, J. Yang, Multi-Attn BLS: multi-head attention mechanism with broad learning system for chaotic time series prediction. Appl. Soft Comput. **132**, 109831 (2023)

59. Y. Dai, Q. Zhou, M. Leng, X. Yang, Y. Wang, Improving the Bi-LSTM model with XGBoost and attention mechanism: a combined approach for short-term power load prediction. Appl. Soft Comput. **130**, 109632 (2022)

Chapter 2
Artificial Intelligence (AI) Applications in Job Sequencing and Scheduling

2.1 Artificial Intelligence (AI)

2.1.1 Taxonomy of AI Technologies

According to Perico and Mattioli [1], AI technologies can be divided into two categories [2]:

- **Data-driven AI** (i.e., brain-style learning) technologies, including artificial neural networks (ANNs), machine learning (ML) (supervised learning, unsupervised learning, statistical learning), deep learning (DL), evolutionary computing, fuzzy logic, etc. Applications of data-driven AI technologies often occur in pattern recognition, classification, clustering, and perception.
- **Symbolic AI** (i.e., modeling and knowledge reasoning) technologies, including ontology, semantic graphs, knowledge-based systems, reasoning, etc. Multi-criteria decision-making (MCDM), production planning, and job scheduling are common applications of symbolic AI technologies.

Figure 2.1 illustrates the classification of AI technologies, which also emphasizes the connection between AI technologies and job scheduling applications. However, this figure is not binding as data-driven AI technologies can also be applied to job scheduling.

AI technologies have been widely used in manufacturing. So far, the application of AI in manufacturing has brought closer connections between people, machines, and computer systems, enabling manufacturers to better optimize processes and solve problems [3]. However, AI technologies applied in manufacturing are naturally different from those applied in other fields. In the manufacturing environment, AI technologies are applied to jobs (such as job scheduling and defect detection) and machines (such as predictive maintenance), while in other fields, AI technologies mainly serve people (such as disease diagnosis and recommendation systems).

© The Author(s), under exclusive license to Springer Nature Switzerland AG 2025
T. T. Chen, *Explainable and Customizable Job Sequencing and Scheduling*,
SpringerBriefs in Applied Sciences and Technology,
https://doi.org/10.1007/978-3-031-85374-6_2

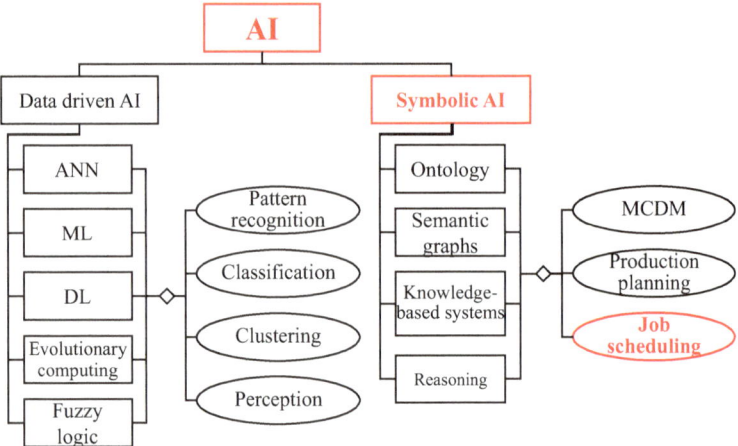

Fig. 2.1 Classification of AI technologies

Some researchers equate Industry 4.0 with AI [4]. The concept of AI 2.0 has also been coined with inconsistent definitions [5–7]. The main features of AI 2.0 include deep learning, Internet-based AI, augmented intelligence, cross-media reasoning, and others [8].

2.1.2   Generative AI

The popular **generative AI** technologies (GenAI) are AI technologies that can create original content (such as text, images, video, audio, or software code) based on user prompts or requests [9].

GenAI relies on DL models to simulate the learning and decision-making processes of the human brain. These models work by identifying and encoding patterns and relationships in large amounts of data, then using that information to understand a user's natural language request or question and respond with relevant new content (see Fig. 2.2).

However, users are still required to complete these pieces. Therefore, generative AI can be viewed as a productivity tool designed to serve people.

GenAI applications are a challenging task in manufacturing:

- First, there are very few GenAI systems specifically designed for users in manufacturing systems.
- In addition, the main purposes of existing GenAI systems (such as information collection and organization, generation of images, videos, or music) do not fall within the functions and requirements of manufacturing systems.
- For factory personnel, existing AI systems with fixed inputs and limited options may be more convenient and easier to use than GenAI systems.

Fig. 2.2 Concept of generative AI

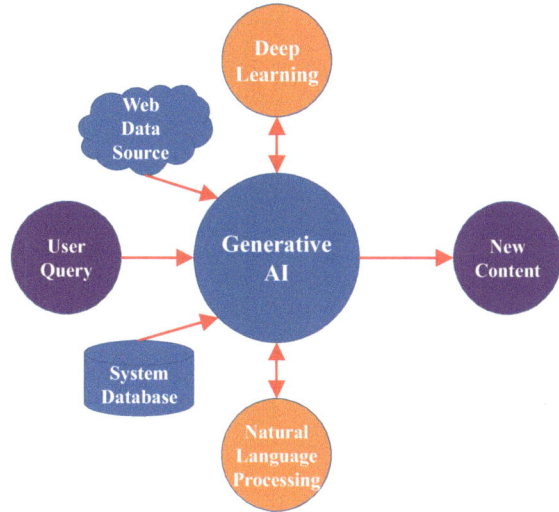

Nonetheless, GenAI promises to promote flexibility, communicability, under-standability, and acceptance of AI applications in manufacturing systems.

2.2   AI Applications in Job Sequencing and Scheduling

2.2.1   Classification of Existing Scheduling Methods

Job scheduling is a basic and important task for every manufacturing system. Regard-less of the size of the manufacturing system, scheduling jobs currently in and to be released to the manufacturing system provides the basis for planning production, transportation, and other supporting activities [10]. Gupta and Sivakumar [11] clas-sified existing scheduling methods into four categories: scheduling heuristics/priority rules, mathematical programming (MP) models, neighborhood search methods, and AI techniques, as shown in Fig. 2.3.

According to Fig. 2.3,

- Traditional AI techniques for solving job scheduling problems include artificial neural networks (ANNs), fuzzy logic, and Petri net-based approaches.
- In addition, AI technologies can also be applied to build a job scheduling system or prepare the inputs required by the system [12, 13].
- Furthermore, for complex job scheduling problems, mathematical programming (MP) models are usually formulated and optimized [11, 14]. However, some of these MP models are intractable (NP hand). As a result, AI methods, such as fuzzy logic, bio-inspired algorithms, and ANNs have been extensively applied to help optimize these MP models (see Fig. 2.3) [12, 13, 15, 16].

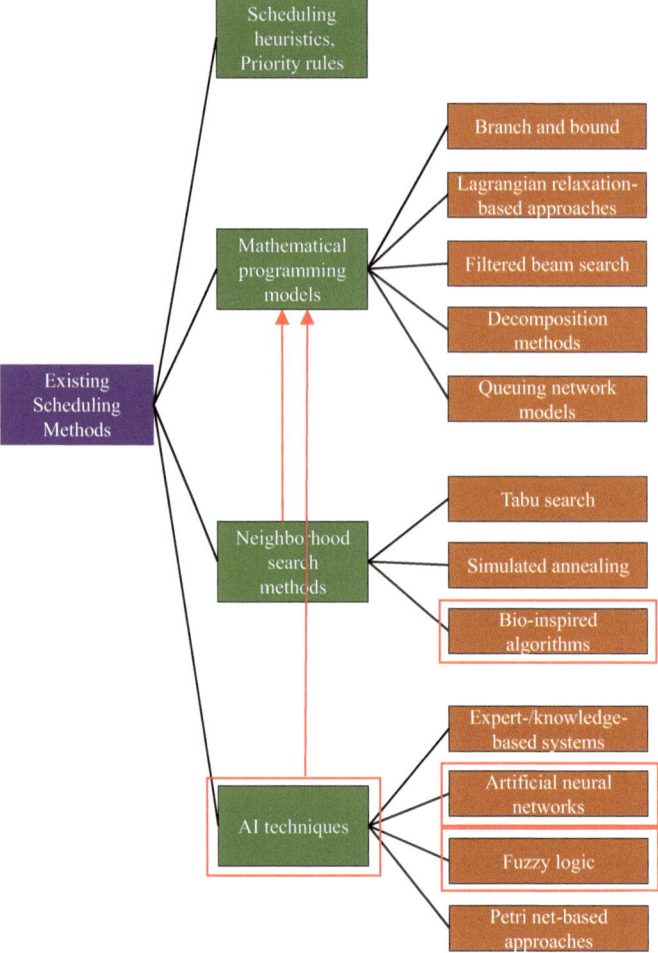

Fig. 2.3 Classification of existing scheduling methods

- Some AI techniques, such as ANNs and genetic algorithms (GA), can also be applied to analyze scheduling results to adjust/optimize scheduling rules [13, 17–19].

2.2.2 Procedure for Applying AI Technologies to Job Sequencing and Scheduling

To select suitable AI technologies, the entire job scheduling process needs to be broken down into the following tasks [20]:

Step 1. Select scheduling performance measures: Common performance measures include makespan (C_{max}), maximum lateness (L_{max}), total completion time ($\sum C_j$), total weighted completion time ($\sum w_j C_j$), total (weighted) tardiness ($\sum (w_j) T_j$), (weighted) number of tardy jobs $\sum (w_j) U_j$, mean cycle time ($\overline{C}$), and cycle time standard deviation (s_C) [20].

Step 2. Collect (or estimate) the required data (see Table 2.1): Some of the data required for job scheduling can be extracted from the production plan and production management information systems [21]. For other data that need to be exported, estimated, or predicted, AI technologies are also suitable [22].

Step 3. Establish (or reestablish) the scheduling model: Job scheduling problems are usually solved by formulating and optimizing a mathematical programming (MP) model [15] or applying dispatching rules [19].

Step 4. Solve the scheduling problem to generate the optimal scheduling plan: AI technologies can be used to find optimal solutions to MP models, prepare inputs for dispatching rules, or optimize dispatching rules [23–26].

Step 5. Implement the optimal scheduling plan.

Step 6. Evaluate the scheduling performance.

Step 7. Return to Step 3.

Table 2.1 Data required for job scheduling [21]

Data type	Collected data	Derived data	Estimated/forecasted data
Examples	• Arrival time • Availability • Costs • Due date • Energy consumption • Job no • Lateness penalty • Lot size • Output time • Price • Priority • Processing time • Product type • Release time • Repair time • Time between failure • Etc.	• Bottleneck • Hourly capacity of following machine • Mean time between failures • Mean time to repair • Minimum and maximum raw/derived data • Next bottleneck • Number of visits • Processing time until next bottleneck • Processing time until next stop (by product type) • Product yield • Remaining processing time (by product type) • Stage cycle time • Utilization of following machine • WIP level in queue of following machine • Etc.	• Cycle time (by product type) • Demand • Energy consumption • Minimum and maximum estimated/forecasted data • Product yield • Remaining cycle time (by-product type) • Time to next failure • Unit cost • Etc.

Fig. 2.4 AI technology
applications to job
scheduling

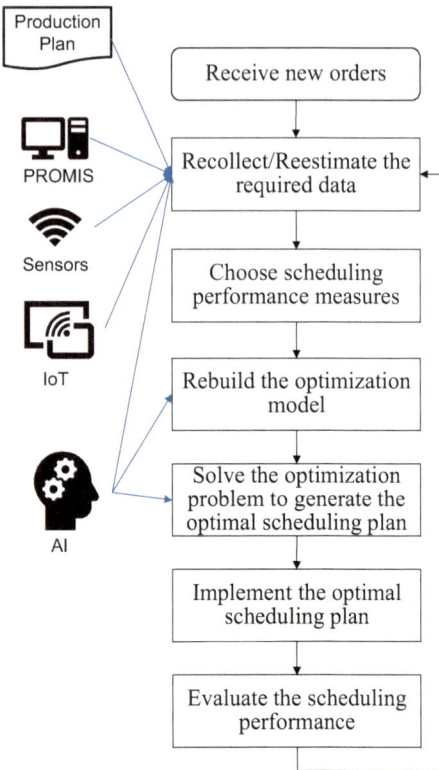

These tasks form an iterative process. As illustrated in Fig. 2.4, AI technologies can be applied in several steps to facilitate job scheduling, and the performance achieved by these AI techniques often far exceeds human capabilities [27].

2.3 Examples of AI Applications in Job Sequencing and Scheduling

2.3.1 ANN

Some examples are given as follows. Wang et al. [28] proposed a fuzzy neural approach to optimize the job scheduling performance in a wafer fabrication factory. They first applied a hybrid **fuzzy c-means (FCM)** and **backpropagation network (BPN)** method to improve the effectiveness of estimating the remaining cycle time of a job. Subsequently, they established a systematic program to determine the optimal values of parameters in a two-factor tailored nonlinear fluctuation smoothing rule for average cycle time to optimize the scheduling performance.

Zhu et al. [29] distributed tasks among multiple autonomous underwater vehicles (AUVs) to balance their workloads. To this end, a **self-organizing map (SOM)** was constructed. Then, they proposed a velocity synthesis method to plan the shortest path for each AUV. Although both objectives belonged to AUVs, the two objectives were optimized sequentially.

Chen and Wang [30] proposed a nonlinear scheduling rule to improve the scheduling performance of semiconductor manufacturing plants by combining a fuzzy neural remaining cycle time estimator. The scheduling rule was modified from the well-known fluctuation smoothing rule by incorporating three novel treatments. First, the remaining cycle time of each job in the semiconductor manufacturing plant was estimated using the look-ahead **SOM-fuzzy backpropagation network (FBPN)** method. Subsequently, both the job release time and the remaining cycle time were normalized to balance their importance in the fluctuation smoothing rule. Finally, the normalized release time was divided by the normalized remaining cycle time to obtain the slack value. In this way, the scheduling rule became nonlinear.

2.3.2 Simulation, Digital Twin

Simulation is a useful tool for evaluating the quality of job scheduling plans. AbuKhousa et al. [31] reviewed several practices for applying simulation to assist healthcare supply chain job scheduling, including scheduling the distribution of scarce drugs (taking into account constraints among clinics, pharmaceutical companies, and distribution), scheduling drug replenishment by considering demand changes and various modalities of drug stocks, and scheduling sterilization logistics in deterministic and non-deterministic scenarios. Although simulation can model the operational details of executing a job scheduling plan, many assumptions need to be made, so simulation results may differ significantly from reality.

The effectiveness of job scheduling systems is usually evaluated through simulation experiments. Chen [32] fuzzified the fluctuation smoothing policy for mean cycle time (FSMCT) and the fluctuation smoothing policy for variation of cycle time (FSVCT) by considering the uncertainty of the remaining cycle time of the job. In this way, the slack overlap problem was solved in a non-subjective way. In addition, a new fuzzy backpropagation network method was also proposed to estimate the remaining cycle time. The performances of the fuzzy scheduling rules were then evaluated through a series of production simulation experiments.

Chen and Wang [33] proposed a nonlinear fluctuation smoothing rule to further improve the performance of job scheduling in a wafer fabrication factory (wafer fab). Two nonlinear forms of the existing fluctuation smoothing rules were derived to enhance the balance and responsiveness, and then they were merged into a bi-criteria rule for considering two performance measures (average cycle time and cycle time variation) at the same time. The content of the bi-criteria rule was tailored for the target wafer fab with an adjustable factor. They also estimated the remaining cycle time of a job by applying the SOM-FBPN approach to improve the estimation accuracy. To evaluate the effectiveness of the proposed methodology, a production simulation was conducted.

Zhang et al. [34] built a five-dimensional digital twin model for a machine in a job shop and then applied the digital twin model to predict the availability of the machine, detect possible disturbance, and evaluate the performance of the machine. The machine scheduling problem was formulated as a nonlinear programming (NLP) problem, but could be converted into a linear programming (LP) problem. The only challenge in solving the scheduling problem was estimating the parameter values in the model that were subject to uncertainty. After collecting such information, the digital twin model was used to perform production simulation to generate a scheduling plan for the machine.

Fang et al. [35] considered a double-resource flexible job shop scheduling problem, in which three jobs were to be processed by five machines with four workers. Each job had three operations with precedence constraints and could be processed by any machine with the assistance of any worker. They formulated such a scheduling problem as a five-objective NLP problem that could be converted into a simpler LP problem. The non-dominated sorting genetic algorithm II (NSGA-II) was applied to help solve the scheduling problem. The execution of the scheduling plan was simulated using a digital twin of the double-resource flexible job shop.

2.3.3   Genetic Algorithm

Several past studies formulated **MP** models for job scheduling. Some of these MP models were not easy to solve. Therefore, heuristics (algorithms) or even **genetic algorithms (GA)** need to be proposed to help solve these problems. Goodarzian et al. [36] formulated a multi-objective mixed integer-nonlinear programming (MINLP) model to schedule the transportation in a multi-layer pharmaceutical supply chain to minimize the total cost and CO_2 emissions while maximizing the shelf life of pharmaceuticals. By converting the MINLP problem into a linear problem, it became more tractable.

Qing-dao-er-ji and Wang [37] proposed a new hybrid GA to solve a job shop scheduling problem, in which a hybrid selection operator based on the fitness value and the concentration value was used. Additionally, a new crossover operator was defined per machine. The mutation operator took into account operations on the critical path. They also employed a local search mechanism to improve the capability of the GA.

De Giovanni and Pezzella [38] considered a distributed flexible job shop scheduling problem (DFJS) consisting of multiple flexible manufacturing units (FMUs). The goal was to minimize the global completion time for all FMUs. They proposed a GA to solve the DFJS problem, where the genetic encoding contained information about allocating jobs to FMUs. Furthermore, the greedy decoding process exploited flexibility and determined job routing. Similar to Qing-dao-er-ji and Wang [32], they also adopted a local search mechanism to improve the available solutions by refining the most promising solutions of each generation.

Davis [39] applied a GA to deal with non-deterministic problems using multiple abstraction levels and progressive constraint relaxation in a framework-based representation system to explain the scheduling results of a job shop scheduling system.

Salido et al. [40] proposed a GA to solve a job shop scheduling problem, where machines consumed different energy to process tasks at different rates (speed scaling). They did not formulate the shop scheduling problem as a mathematical programming (MP) problem. As long as a special fitness function that takes energy consumption into account is defined, such a job shop scheduling problem can be solved directly using GAs.

Pezzella et al. [41] scheduled the operations of three jobs on four machines. The manufacturing system was a flexible job shop where the numbers of operations for different jobs and the processing times of other jobs on the same machine differed. Additionally, each operation can be performed on multiple machines with varying processing times. Optimizing the makespan of such a manufacturing system was an NP-hard problem. Theoretically, the number of possible permutations of this problem was at most $4^{3+3+2} \cdot 8! = 2.64 \cdot 10^9$. Therefore, a **GA** was applied to help solve the job scheduling problem.

Seman et al. [42] formulated the nanosatellite task scheduling problem as a mixed integer-linear programming (MILP) problem and proposed a hybrid GA-dynamic programming to help solve the MILP problem. Following the concept of locally interpretable model-agnostic explanations (LIME) [43], a linear regression (LR) equation between the decision variable values and the objective function value was fitted around the optimal solution to provide a local explanation.

2.3.4 Industry 4.0

Chen and Wang [44] established an advanced **Internet of things (IoT)** system to assist ubiquitous manufacturing of multiple **three-dimensional (3D) printing** facilities. The system received customer orders online and then distributed the required parts to nearby 3D printing facilities. After printing was completed, a freight truck proceeded to the 3D printing facilities to collect the prints. To minimize the cycle time to deliver an order, a MILP model was first optimized to achieve workload balancing, and then a mixed integer-quadratic programming (MIQP) problem was solved to find the shortest delivery path. Both models were NP-hard. Therefore, two **branch-and-bound (B&B)** algorithms were used to facilitate the search for the global optimal solutions.

Kianpur et al. [45] applied Industry 4.0 technologies to collect real-time information about processing times (including unexpected events) and due dates to generate dynamic and adaptive schedules for a job shop to improve the scheduling performance. A MILP model was formulated to minimize earliness and tardiness costs while taking into account rescheduling costs, for which scheduling heuristics such as earned value (EV) and predicted total cost at completion (EACf) were applied.

Li et al. [46] solved a flexible job shop scheduling problem with sequence-dependent settings and limited dual resources. A MILP model was formulated for this purpose. Then, a hybrid metaheuristic approach consisting of a genetic algorithm (GA) and Tabu search (TS) was adopted to solve the MILP problem. To avoid frequent rescheduling, three ML methods (random forest, SVM, and multi-layer perceptron) were used to identify rescheduling patterns (i.e., changes in production conditions worth considering to be rescheduled).

2.3.5 Fuzzy Logic

Chiu and Chen [47] established a healthcare supply chain with multiple 3D printing facilities to manufacture dentures. The healthcare supply chain was also working with home delivery services, which have become increasingly popular during the COVID-19 pandemic, to save on shipping time. In addition, interval type-II trapezoidal fuzzy numbers were used to model the printing and shipping times and consider their uncertainties. An interval type-II fuzzy mixed integer-linear programming (FMILP) model was formulated and optimized for scheduling the operations in the healthcare supply chain. **Fuzzy logic** has shown its effectiveness in modeling inevitable operational uncertainties in manufacturing systems.

Demirli and Yimer [48] proposed an integrated production–distribution planning model for a multi-echelon, multi-plant and multi-product build-to-order (BTO) supply chain. The uncertainties associated with various operational costs were represented by fuzzy numbers. Then, a FMILP model to minimize the overall operating costs of fabrication, procurement, assembling, inspection, logistics, and inventory was formulated, while meeting customer satisfaction thresholds and due dates at each outlet. They also proposed a compromise solution approach to convert the FMILP problem into an auxiliary multi-objective linear programming model to facilitate problem-solving.

2.3.6 Ant Colony Optimization, Invasive Weed Optimization

Ant colony optimization (ACO) has revealed its effectiveness (or potential) in solving various job scheduling problems. For example, Udhayakumar and Kumanan [49] attempted to minimize the makespan of a flexible manufacturing system (FMS) using ACO, where job operations can be performed on any of multiple machines with varying processing times.

Li et al. [50] applied ACO to help solve an identical parallel batch processing machine (PBPM) scheduling problem, where multiple jobs (called batches) can be processed together on any of multiple identical machines, as is typical in wafer fabrication. Minimizing the average cycle time is a common goal in scheduling such systems.

For a similar problem, Zhang et al. [51] considered the situation where various tools were shared between machines and each operation could be processed using different tools. They proposed a **max–min ACO** algorithm, in which the pheromones along a path were restricted to a range, for which the upper and lower boundaries were updated separately at each iteration.

Wu et al. [52] applied ACO to solve a multi-objective flexible job shop scheduling problem. The objective functions were compromised by forming their weighted sum to be minimized. However, such a treatment may be problematic since the objective functions had different units and ranges.

Ghasemi et al. [53] formulated a multi-objective MILP model to schedule the transportation and inventory replenishment of a plasma supply chain to maximize donor coverage during periods while minimizing transportation, relocation, inventory holding, and establishment costs. They applied **the ε-constraint method** and **the multi-objective invasive weed optimization (MOIWO) algorithm** to solve the multi-objective MILP problem when the supply chain scale was small and large, respectively.

2.4 Problems with Existing AI Applications

The existing applications of AI technologies in job sequencing and scheduling have the following problems:

- Many personnel related to job scheduling lack background knowledge of AI technologies, which hinders their understanding, communication, and acceptance of the applications of AI technologies.
- At the same time, job scheduling methods incorporating more complex AI technologies are being proposed. It has become increasingly difficult to examine the rationality and correctness of the scheduling mechanism and the optimality of the scheduling results.
- In order to improve the understandability and credibility of these AI applications in job scheduling and ensure the traceability, explainability, and accountability of job scheduling results, explainable artificial intelligence (XAI) is applicable. However, the applications of XAI in this field are still lacking [3].

2.5   Generative AI Applications in Job Sequencing and Scheduling

Generative AI applications are a challenging task in job scheduling:

- First, there are no generative AI systems specifically designed for this purpose.
- Furthermore, existing generative AI systems may not be directly applicable to job scheduling, because it has been difficult to input the data and constraints required for scheduling into existing generative AI systems so far.

Job scheduling in factories of considerable size is often accomplished using computerized systems, which interfaces often limit the queries that can be responded to, not to mention in a natural language way. Nonetheless, generative AI promises to promote flexibility, communicability, understandability, and acceptance of job scheduling. The input interface of a GenAI-based job scheduling system is shown in Fig. 2.5.

Nevertheless, compared to other fields, job scheduling is a simpler field for gen AI applications for the following reasons:

- The purpose is relatively simple (only used for job scheduling-related purposes).
- As a result, queries contain only limited keywords, making the natural language processing (NLP) task more straightforward.
- There are much fewer users.
- Users can express their needs in the same language after long-term communication.

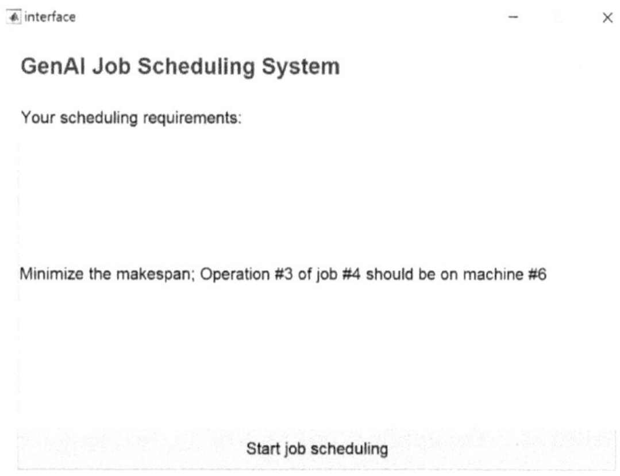

Fig. 2.5 Input interface of GenAI-based job scheduling system

```
▲ interface                                              —      ×

GenAI Job Scheduling System

Your scheduling requirements:




Minimize the makespan; Operation #3 of job #4 should be on machine #6




                        Start job scheduling
```

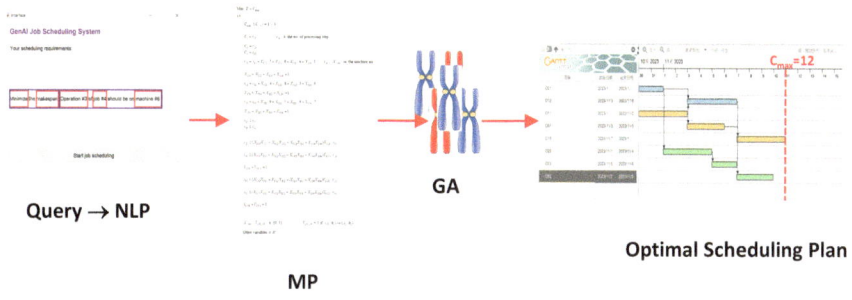

Fig. 2.6 Procedure for applying GenAI to job scheduling

The procedure for applying GenAI to job scheduling may include the following steps (see Fig. 2.6).

Step 1. The user enters a query.
Step 2. The system processes the query using NLP techniques.
Step 3. The system formulates the corresponding scheduling (e.g., MP) problem.
Step 4. The system applies AI (e.g., GA) to solve the scheduling problem.
Step 5. The system responds to the user with the scheduling plan.

References

1. P. Perico, J. Mattioli, Empowering process and control in lean 4.0 with artificial intelligence, in *Third International Conference on Artificial Intelligence for Industries* (2020), pp. 6–9
2. T.-C.T. Chen, Y.-C. Wang, Artificial intelligence in manufacturing, in *Artificial Intelligence and Lean Manufacturing* (2022), pp. 13–35
3. Y.C. Wang, T. Chen, Adapted techniques of explainable artificial intelligence for explaining genetic algorithms on the example of job scheduling. Expert. Syst. Appl. **237**(A), 121369 (2024)
4. D. Devereaux, Smaller manufacturers get lean with artificial intelligence (2019). www.nist.gov/blogs/manufacturing-innovation-blog/smaller-manufacturers-get-leanartificial-intelligence
5. P. Palensky, D. Bruckner, A. Tmej, T. Deutsch, Paradox in AI–AI 2.0: the way to machine consciousness, in *International Conference on IT Revolutions* (2008), pp. 194–215
6. Y.H. Pan, Heading toward artificial intelligence 2.0. Engineering **2**(4), 409–413 (2016)
7. P.J. Lisboa, (2018). AI 2.0: augmented intelligence, data science and knowledge engineering for sensing decision support, in *Proceedings of the 13th International FLINS Conference* (2018), pp. 10–11
8. C. Stryker, M. Scapicchio, What is generative AI? (2024). https://www.ibm.com/topics/generative-ai
9. B.H. Li, B.C. Hou, W.T. Yu, X.B. Lu, C.W. Yang, Applications of artificial intelligence in intelligent manufacturing: a review. Front. Inf. Technol. Electron. Eng. **18**(1), 86–96 (2017)
10. Y.C. Wang, T. Chen, C.W. Lin, A slack-diversifying nonlinear fluctuation smoothing rule for job dispatching in a wafer fabrication factory. Robot. Comput.-Integr. Manuf. **29**(3), 41–47 (2013)
11. A.K. Gupta, A.I. Sivakumar, Job shop scheduling techniques in semiconductor manufacturing. Int. J. Adv. Manuf. Technol. **27**, 1163–1169 (2006)

12. H. Wang, Flexible flow shop scheduling: optimum, heuristics and artificial intelligence solutions. Expert. Syst. **22**(2), 78–85 (2005)
13. T. Chen, Y.-C. Wang, A fuzzy-neural approach for supporting three-objective job scheduling in a wafer fabrication factory. Neural Comput. Appl. **23**(1), 353–367 (2013)
14. M. Del Gallo, G. Mazzuto, F.E. Ciarapica, M. Bevilacqua, Artificial intelligence to solve production scheduling problems in real industrial settings: systematic literature review. Electronics **12**(23), 4732 (2023)
15. G. El Khayat, A. Langevin, D. Riopel, Integrated production and material handling scheduling using mathematical programming and constraint programming. Eur. J. Oper. Res. **175**(3), 1818–1832 (2006)
16. A. Seker, S. Erol, R. Botsali, A neuro-fuzzy model for a new hybrid integrated process planning and scheduling system. Expert Syst. Appl. **40**(13), 5341–5351 (2013)
17. A. Al-Refaie, M. Judeh, T. Chen, Optimal multiple-period scheduling and sequencing of operating room and intensive care unit. Oper. Res. Int. Journal **18**, 645–670 (2018)
18. E.-G. Talbi, Combining metaheuristics with mathematical programming, constraint programming and machine learning. Ann. Oper. Res. **240**(1), 171–215 (2016)
19. T.-C.T. Chen, Job sequencing and scheduling, in *Production Planning and Control in Semiconductor Manufacturing* (2023), pp. 77–100
20. L.P. Michael, *Scheduling: Theory, Algorithms, and Systems* (Springer, 2018)
21. T.-C.T. Chen, Evaluating the sustainability of a smart technology application to mobile health care: the FGM–ACO–FWA approach. Complex Intell. Syst. **6**(1), 109–121 (2020)
22. T. Chen, Job remaining cycle time estimation with a post-classifying fuzzy-neural approach in a wafer fabrication plant: a simulation study. Proc. Inst. Mech. Eng. Part B J. Eng. Manuf. **223**(8), 1021–1031 (2009)
23. R. Babukartik, P. Dhavachelvan, Hybrid Algorithm using the advantage of ACO and Cuckoo Search for Job scheduling. Int. J. Inf. Technol. Converg. Serv. **2**(4), 25–34 (2012)
24. J.-J. Liu, J.-C. Liu, Permeability predictions for tight sandstone reservoir using explainable machine learning and particle swarm optimization. Geofluids **2022**, 2263329 (2022)
25. S.S. Sana, H. Ospina-Mateus, F.G. Arrieta, J.A. Chedid, Application of genetic algorithm to job scheduling under ergonomic constraints in manufacturing industry. J. Ambient. Intell. Humaniz. Comput. **10**(5), 2063–2090 (2019)
26. D. Thiruvady, A.T. Ernst, G. Singh, Parallel ant colony optimization for resource constrained job scheduling. Ann. Oper. Res. **242**(2), 355–372 (2016)
27. A. Holzinger, The next frontier: AI we can really trust, in *Machine Learning and Principles and Practice of Knowledge Discovery in Databases: International Workshops of ECML PKDD 2021, Part I* (2022), pp. 427–440
28. Y.C. Wang, C.W. Wu, T. Chen, A fuzzy-neural approach for optimizing the performance of job dispatching in a wafer fabrication factory. Int. J. Adv. Manuf. Technol. **67**, 189–202 (2013)
29. D. Zhu, H. Huang, S.X. Yang, Dynamic task assignment and path planning of multi-AUV system based on an improved self-organizing map and velocity synthesis method in three-dimensional underwater workspace. IEEE Trans. Cybern. **43**(2), 504–514 (2013)
30. T. Chen, Y.C. Wang, A nonlinear scheduling rule incorporating fuzzy-neural remaining cycle time estimator for scheduling a semiconductor manufacturing factory—a simulation study. Int. J. Adv. Manuf. Technol. **45**, 110–121 (2009)
31. E. AbuKhousa, J. Al-Jaroodi, S. Lazarova-Molnar, N. Mohamed, Simulation and modeling efforts to support decision making in healthcare supply chain management. Sci. World J. **2014**, 354246 (2014)
32. T. Chen, A fuzzy rule for job dispatching in a wafer fabrication factory—a simulation study. Int. J. Adv. Manuf. Technol. **67**, 47–58 (2013)
33. T. Chen, Y.-C. Wang, A bi-criteria nonlinear fluctuation smoothing rule incorporating the SOM-FBPN remaining cycle time estimator for scheduling a wafer fab—a simulation study. Int. J. Adv. Manuf. Technol. **49**(5), 709–721 (2010)
34. M. Zhang, F. Tao, A.Y.C. Nee, Digital twin enhanced dynamic job-shop scheduling. J. Manuf. Syst. **58**, 146–156 (2021)

35. Y. Fang, C. Peng, P. Lou, Z. Zhou, J. Hu, J. Yan, Digital-twin-based job shop scheduling toward smart manufacturing. IEEE Trans. Industr. Inf. **15**(12), 6425–6435 (2019)
36. F. Goodarzian, P. Ghasemi, A. Appolloni, I. Ali, L.E. Cárdenas-Barrón, Supply chain network design based on big data analytics: heuristic-simulation method in a pharmaceutical case study. Prod. Plan. Control. 1–21 (2024)
37. R. Qing-dao-er-ji, Y. Wang, A new hybrid genetic algorithm for job shop scheduling problem. Comput. Oper. Res. **39**(10), 2291–2299 (2012)
38. L. De Giovanni, F. Pezzella, An improved genetic algorithm for the distributed and flexible job-shop scheduling problem. Eur. J. Oper. Res. **200**(2), 395–408 (2010)
39. L. Davis, Job shop scheduling with genetic algorithms, in *Proceedings of the first International Conference on Genetic Algorithms and their Applications* (2014), pp. 136–140
40. M.A. Salido, J. Escamilla, A. Giret, F. Barber, A genetic algorithm for energy-efficiency in job-shop scheduling. Int. J. Adv. Manuf. Technol. **85**, 1303–1314 (2016)
41. F. Pezzella, G. Morganti, G. Ciaschetti, A genetic algorithm for the flexible job-shop scheduling problem. Comput. Oper. Res. **35**(10), 3202–3212 (2008)
42. L.O. Seman, C.A. Rigo, E. Camponogara, E.A., Bezerra, L. dos Santos Coelho, Explainable column-generation-based genetic algorithm for knapsack-like energy aware nanosatellite task scheduling. Appl. Soft Comput. 110475 (2023)
43. P. Nagaraj, V. Muneeswaran, A. Dharanidharan, K. Balananthanan, M. Arunkumar, C. Rajkumar, A prediction and recommendation system for diabetes mellitus using XAI-based lime explainer, in *2022 International Conference on Sustainable Computing and Data Communication Systems* (2022), pp. 1472–1478
44. T. Chen, Y.C. Wang, An advanced IoT system for assisting ubiquitous manufacturing with 3D printing. Int. J. Adv. Manuf. Technol. **103**, 1721–1733 (2019)
45. P. Kianpour, D. Gupta, K.K. Krishnan, B. Gopalakrishnan, Automated job shop scheduling with dynamic processing times and due dates using project management and industry 4.0. J. Ind. Prod. Eng. **38**(7), 485–498 (2021)
46. Y. Li, S. Carabelli, E. Fadda, D. Manerba, R. Tadei, O. Terzo, Machine learning and optimization for production rescheduling in Industry 4.0. Int. J. Adv. Manuf. Technol. **110**(9), 2445–2463 (2020)
47. M.C. Chiu, T.C.T. Chen, A ubiquitous healthcare system of 3D printing facilities for making dentures: application of type-II fuzzy logic. Digit. Health **34**, 20552076221092540 (2022)
48. K. Demirli, A.D. Yimer, Fuzzy scheduling of a build-to-order supply chain. Int. J. Prod. Res. **46**(14), 3931–3958 (2008)
49. P. Udhayakumar, S. Kumanan, Sequencing and scheduling of job and tool in a flexible manufacturing system using ant colony optimization algorithm. Int. J. Adv. Manuf. Technol. **50**, 1075–1084 (2010)
50. L. Li, P. Gu, F. Qiao, Y. Wu, Q. Wu, ACO-based multi-objective scheduling of identical parallel batch processing machines in semiconductor manufacturing, in *Future Manufacturing Systems* (2010), pp. 163–178
51. X. Zhang, S. Wang, L. Yi, H. Xue, S. Yang, X. Xiong, An integrated ant colony optimization algorithm to solve job allocating and tool scheduling problem. Proc. Inst. Mech. Eng. Part B J. Eng. Manuf. **232**(1), 172–182 (2018)
52. J. Wu, G.D. Wu, J.J. Wang, Flexible job-shop scheduling problem based on hybrid ACO algorithm. Int. J. Simul. Model. **16**(3), 497–505 (2017)
53. P. Ghasemi, F. Goodarzian, V. Simic, E.B. Tirkolaee, A DEA-based simulation-optimisation approach to design a resilience plasma supply chain network: a case study of the COVID-19 outbreak. Int. J. Syst. Sci. Oper. Logist. **10**(1), 2224105 (2023)

Chapter 3
XAI Applications in Job Sequencing and Scheduling

3.1 XAI Applications for Collecting and Estimating the Required Data

This chapter introduces the application of explainable artificial intelligence (XAI) according to the process of job sequencing and scheduling as shown in Fig. 3.1. First, various artificial intelligence (AI) technologies have been applied to collect and/or estimate the required data for job sequencing and scheduling, so there is an urgent need to use XAI techniques and tools to explain such AI applications.

For example, Chen and Wang [1] proposed the nonlinear fluctuation smoothing policy for variance of cycle time (FSVCT):

$$\mathrm{SK}_{jk} = \frac{R_j - \min_l R_l}{\max_l R_l - \min_l R_l} \bigg/ \frac{\mathrm{RCTE}_{jk} - \min_l \mathrm{RCTE}_{lk}}{\max_l \mathrm{RCTE}_{lk} - \min_l \mathrm{RCTE}_{lk}}, \tag{3.1}$$

where SK_{jk} represents the slack of job j with operation k to be performed; $j = 1 \sim n$, $k = 1 \sim K$. The job with the smallest slack will be processed first [2]. R_j is the release time of this job, while RCTE_{jk} is its estimated remaining cycle time:

$$\mathrm{RCTE}_{jk} = (\mathrm{CTE}_j - \mathrm{SCT}_{jk}) \cdot \frac{\mathrm{SCT}_{jk}}{\mathrm{SCTE}_{jk}}, \tag{3.2}$$

where CTE_j, SCT_{jk}, and SCTE_{jk} denote the estimated cycle time, step cycle time, and estimated step cycle time of this job, respectively. Both CTE_j and SCTE_{jk} need to be estimated. RCTE_{jk} is also an important information for scheduling rules such as critical ratio (CR) [3] and fluctuation smoothing policy for mean of cycle time (FSMCT) [4]. To this end, they applied the look-ahead self-organization map (SOM)-fuzzy back-propagation network (FBPN) approach [5], in which jobs were classified using a SOM [6, 7] before estimating their CTE_j (or SCTE_{jk}) using a FBPN.

© The Author(s), under exclusive license to Springer Nature Switzerland AG 2025
T. T. Chen, *Explainable and Customizable Job Sequencing and Scheduling*,
SpringerBriefs in Applied Sciences and Technology,
https://doi.org/10.1007/978-3-031-85374-6_3

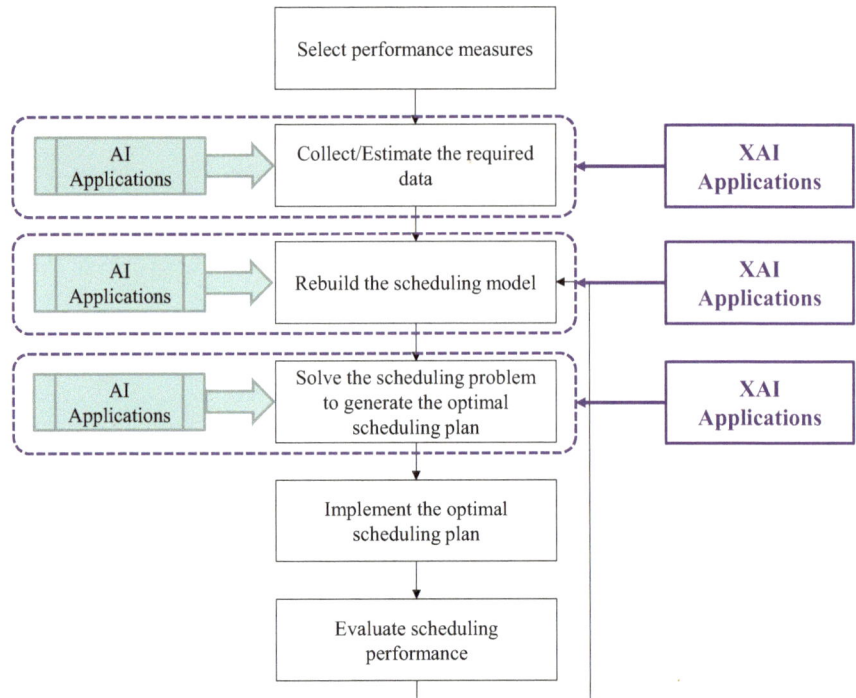

Fig. 3.1 XAI applications according to the process of job sequencing and scheduling

The process and results of high-dimensional classification using SOM or other clas-sification methods are unintuitive and difficult to visualize, leading to difficulties in understanding [8]. In addition, fuzzy neural networks (FNNs) are more difficult to understand or explain than artificial neural networks (ANNs) that have been consid-ered as black boxes [9]. To date, there have been few attempts to apply XAI to general fuzzy systems, rule-based fuzzy systems, and neuro-fuzzy systems [10, 11].

3.2 Explaining AI Application for Estimating the Required Data

3.2.1 Visualization XAI Techniques and Tools

Traditionally, visualization techniques and tools have been widely used to illustrate scheduling processes/mechanisms and results. The effectiveness of such applications demonstrates the potential of visualization XAI techniques and tools to explain AI applications in job sequencing and scheduling.

The first visualization technique is **system (architecture) diagrams**, which is also the most popular method for explaining similar AI applications. For example, Chen [5] drew a system (architecture) diagram (see Fig. 3.2) to illustrate the SOM-FBPN method for estimating the remaining cycle time of a job. In XAI terms, such methods aim to provide **global explanations**.

In addition to showing the system architecture, some XAI tools (e.g., R, ConvNetJS, MATLAB, TensorFlow, Seq2Seq, etc.) also visualize the operations of an ANN application for similar purposes. Such visualization XAI tools are expected to [12]

- show the network architecture,
- show the values of network parameters,
- show the convergence process of network parameters, and
- show the convergence process of results.

Chen [12] divided the existing XAI techniques for explaining AI applications in collecting and estimating the required data for job sequencing and scheduling into the following categories:

- XAI techniques for assessing the contribution, influence, effect, importance of each input (e.g., job attribute) in predicting the output (e.g., the remaining cycle time).
- XAI techniques for approximating the relationship between inputs and the output, e.g., the trained FBPN, with simpler and more understandable rules.

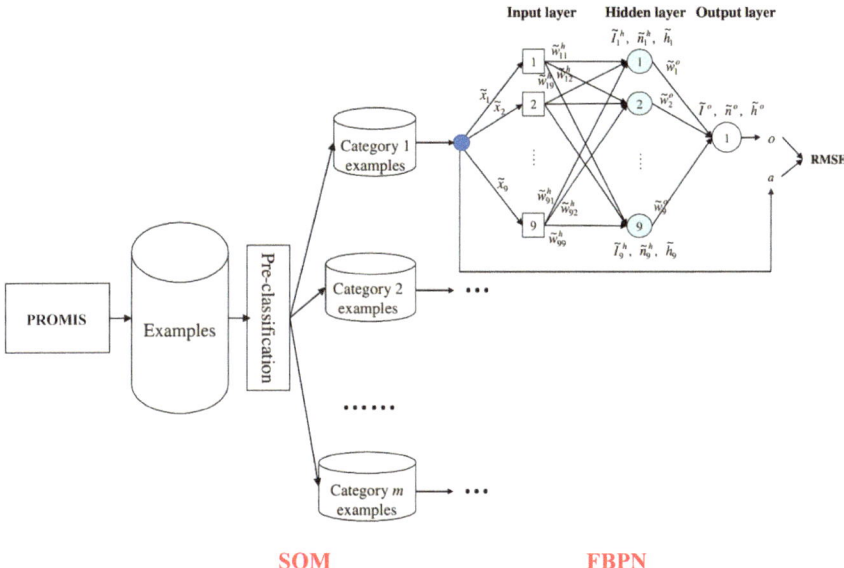

Fig. 3.2 System (architecture) diagram of the SOM-FBPN approach

- XAI techniques to facilitate the understanding of the prediction process (i.e., the operations of the FBPN) and result (e.g., the estimated remaining cycle time).

Visualization XAI techniques help understand the estimation process and fall into the third category. In addition, visualization XAI techniques can also help realize the first category of functions by highlighting the most important parts of the system architecture diagram (see Fig. 3.3). In this figure, the connection from the 2nd and nth inputs to the hidden-layer nodes have higher weights, which means these inputs are more important. In addition, the connection weight from the second hidden-layer node to the output node is also higher than the others, meaning that the hidden-layer node is the most important.

A popular XAI tool for visualizing input distributions is a violin plot. Taking the flexible job shop scheduling problem in Pezzella et al. [13] as an example. The distribution of processing times for various jobs on a machine can be visualized using the violin shown in Fig. 3.4. Obviously,

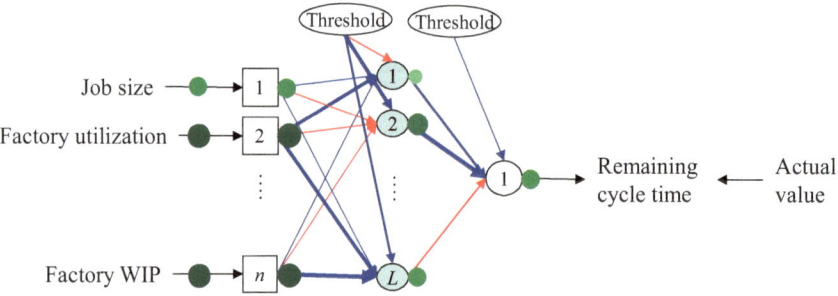

Fig. 3.3 Highlighting the most important parts of the system architecture diagram

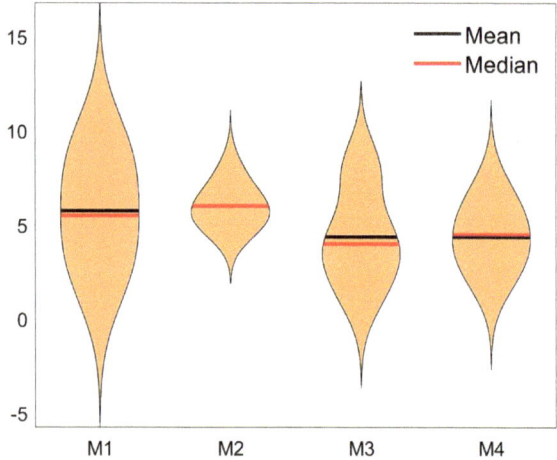

Fig. 3.4 Distribution of processing times for various jobs on the same machine

- Job processing times are more widely distributed on M1.
- In contrast, job processing times on M2 are close.
- In addition, the differences in the average processing times on these machines are very small.

3.2.2 XAI Techniques for Evaluating the Importance of Each Input on the Output

First, there are a variety of ways to analyze the effect of an input on the output (see Fig. 3.5):

- The traditional way is to conduct a **correlation analysis**. For example, depending on the relationship that exists between the inputs and output, the Pearson correlation coefficient [14] or Spearman's rank correlation [15] between each input and the output can be calculated. The input with the highest correlation coefficient is the most influential in determining the output.
- The influence of each input/attribute on the output can also be evaluated by taking the **partial derivative** of the output [16]. The impact of an input is proportional to the result.
- The **odd ratio** of each input can also be calculated for the same purpose:

$$\mathrm{OR}_p = \frac{o_j(x_{jp} + \sigma_p) \cdot (1 - o_j(x_{jp}))}{o_j(x_{jp}) \cdot (1 - o_j(x_{jp} + \sigma_p))}, \tag{3.3}$$

Fig. 3.5 Methods for evaluating the importance of an input to the output

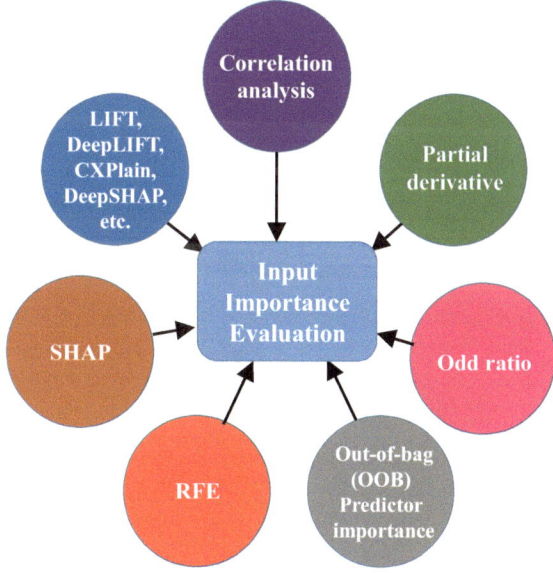

where $o_j(*)$ means the (normalized) output of example j when $x_{jp} = *$.

- **Out-of-bag (OOB) predictor importance** evaluates the importance of an input by observing the change in the estimation error by slightly varying the value of the input:

$$OOB_p = \overline{\varepsilon}_{\cdot p} \big/ s_{\varepsilon_{\cdot p}}, \tag{3.4}$$

where $\overline{\varepsilon}_{\cdot p}$ and $s_{\varepsilon_{\cdot p}}$ are the average and standard deviation of ε_{jp}, respectively; ε_{jp} is the estimation error of example j.

- **Recursive feature elimination (RFE)** can also be used to evaluate the importance of an input by estimating the percentage increase in the estimation error by removing the input.
- **Shapley additive explanations (SHAP)** [17] can be applied simply by evaluating the contribution of an input by fixing its value while randomizing the values of the other inputs.
- **Learning important features (LIFT)** [18, 19]: LIFT is a measure of the effectiveness of an estimation model, calculated as the ratio between the results obtained with and without the estimation model.
- **Deep Learning Important Features (DeepLIFT)** [20]: DeepLIFT decomposes the output prediction of an ANN or deep neural network (DNN) for a specific input by backpropagating the contributions of all neurons in the ANN (or DNN) to each feature of the input. It compares each neuron's activation to its reference activation and assigns a contribution score based on the difference. This allows us to understand the relative impact of each feature on model estimations.
- **Deep SHAP** [21]: DeepSHAP is designed to explain deep learning models by performing DeepLIFT.
- **Causal explanation (CXPlain)** [22]: CXPlain uses a causal objective to train a supervised model to learn to explain another machine learning model. Once trained, CXPlain can interpret a target model in a fraction of the time and is able to quantify the uncertainty associated with its feature importance estimates via bootstrap ensembling.

Inputs to a job scheduling system include job attributes, production conditions, machine statuses, processes, etc. Some input values cannot be changed, let alone randomized. Therefore, some of the above methods cannot be applied. For example,

- You cannot fix the processing time of one job while randomizing those of other jobs, which is practically impossible and meaningless.
- It is also questionable to advance or delay the availability of a machine to observe its impact on the scheduling performance of the optimal solution.

3.2.3 XAI Techniques for Approximating the Estimation Mechanism

Deep learning (DL) applications in this field are often difficult to explain and communicate [23, 24]. Therefore, there are various XAI techniques to explain such DL applications through approximate their reasoning mechanisms (see Fig. 3.6).

First, a **case-based reasoning** (CBR) system can be established to approximate the operations in an ANN (or DNN) for estimating the remaining cycle time [25]. A CBR system is composed of cases $\{((\{r_{kp}|p = 1 \sim P\}, s_k)|k = 1 \sim K\}$ that are used in K rules:

$$\text{"If } x_{j1} = r_{k1} \text{ and } \dots \text{ and } x_{jP} = r_{kP} \text{ then } \hat{o}_j = s_k\text{"}; k = 1 \sim K \qquad (3.5)$$

where $\hat{o}_j$ is the approximated output of the ANN (or DNN) for job j. k-means [26, 27] is usually applied to extract cases from the collected (or synthetic) data. To approximate the output of the ANN (or DNN) for job j, the distance between job j and case k is measured:

$$d_{jk} = \sqrt{\sum_{p=1}^{P} (x_{jp} - r_{kp})^2}. \qquad (3.6)$$

Then, the ANN (DNN) output of this job is approximated as

$$\hat{o}_j = \sum_{k=1}^{K} \left(\frac{\frac{1}{d_{jk}}}{\sum_{l=1}^{K} \frac{1}{d_{jl}}} \cdot s_k \right). \qquad (3.7)$$

Fig. 3.6 XAI techniques for approximating the reasoning mechanisms of DL applications

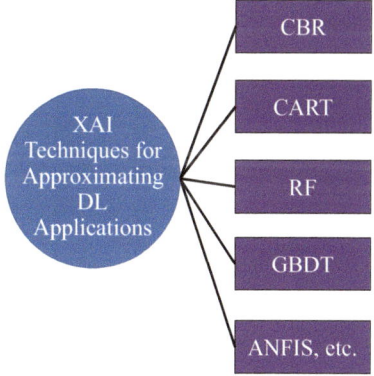

Classification and regression trees (CARTs) are another XAI technique for approximating the operations in the ANN (or DNN) for estimating the remaining cycle time, which has the following form:

$$\text{"If } x_{j(1)} \geq \text{ (or } <)r_{k(1)} \text{ and } \dots \text{ and } x_{j(L)} \geq \text{ (or } <)r_{k(L)} \text{ then } \hat{o}_j = s_k \text{"; } k = 1 \sim K \tag{3.8}$$

An example is provided in Fig. 3.7. It is not necessary to include all attributes in a single rule. In addition, the order in which attributes appear in different rules may not be the same. Furthermore, unlike in CBR, in CART there is no mechanism to aggregate the outcomes of all rules. The procedure for constructing a CART comprises three steps: tree growing, stopping, and pruning [28].

The third XAI technique to approximate the operations in the ANN (or DNN) for estimating the remaining cycle time is **random forests** (RFs). A random forest (RF) is the ensemble of multiple trees that consider different inputs and data parts [29]. An example is given in Fig. 3.8. For a job, a decision rule in each tree can be applied to estimate the remaining cycle time:

$$\text{"If } x_{j(1)}(t) \geq \text{ (or } <)r_{k(1)}(t) \text{ and } \dots \text{ and } x_{j(L)}(t)$$
$$\geq \text{ (or } <)r_{k(L)}(t) \text{ then } \hat{o}_j(t) = s_k(t)\text{"; } t = 1 \sim T \tag{3.9}$$

Then, the estimation results by all trees are averaged:

$$\hat{o}_j = \frac{\sum_{t=1}^{T} \hat{o}_j(t)}{T} \tag{3.10}$$

where t is the index of a tree; $t = 1 \sim T$.

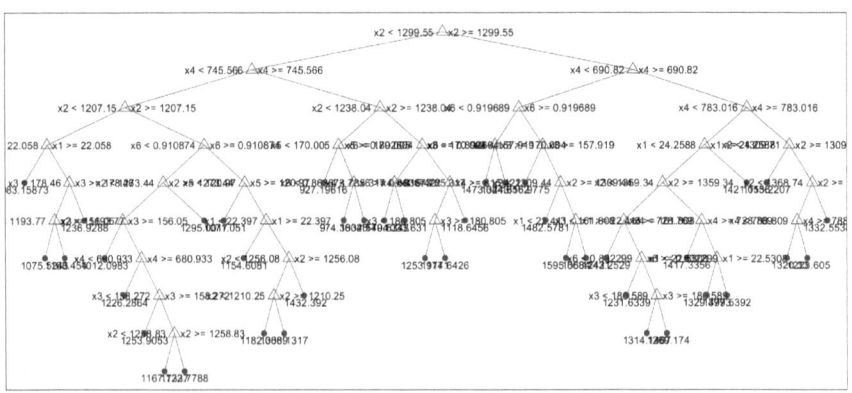

Fig. 3.7 Classification and regression tree for approximating the operations in the ANN for estimating the remaining cycle time

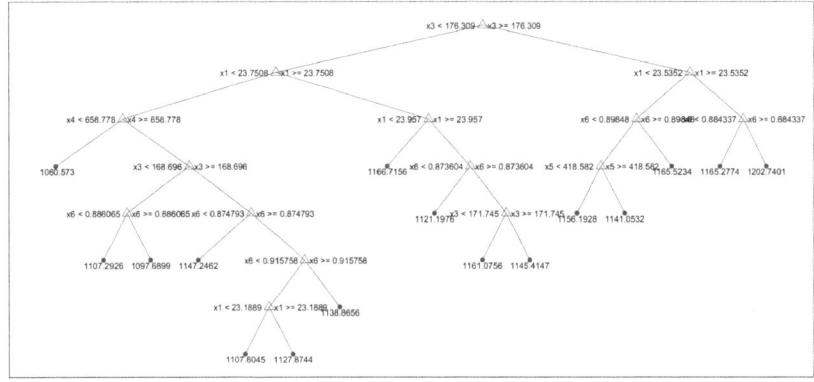

(Tree #1)

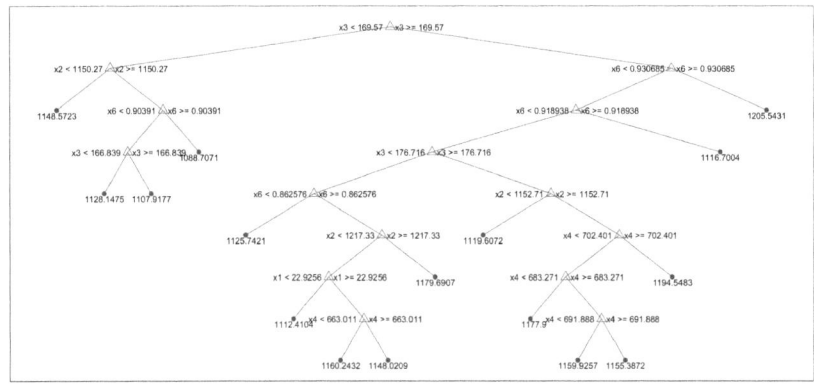

(Tree #2)

...

(Tree #*T*)

Fig. 3.8 RF for approximating the operations in the ANN for estimating the remaining cycle time

The fourth XAI technique for approximating the operations in the ANN (or DNN) for estimating the remaining cycle time is **gradient boosted decision trees** (GBDT) [30], which consist of multiple sequentially trained decision trees. Starting from the first decision tree, each subsequent decision tree improves the previous decision tree by predicting the estimation error of the previous decision tree (the concept of boosting). An example is provided in Fig. 3.9. It is worth noting that the results of Tree #2 are much smaller than those of Tree #1 because the former tries to estimate the estimation error rather than the value of the latter.

By considering different inputs and data parts, a GBDT can be turned into an extreme gradient boosting (XGBoost) decision trees [31].

3.3 XAI Applications for Explaining the Scheduling Mechanism

Dispatch rules are usually defined in the form of linear regression (LR), polynomial, modulo, or simple inverse functions (see Table 3.1), which are easily understandable and do not need explanations [32].

There are some dedicated visualization techniques or tools used for specific AI applications in job sequencing and scheduling, such as chromosome diagrams for GA and disjunction diagrams for ant colony optimization (ACO). Except for these dedicated techniques or tools, many generic XAI techniques and tools are applicable for explaining the scheduling mechanisms of various AI applications in job sequencing and scheduling.

3.3.1 Generic XAI Techniques and Tools

Textual descriptions are probably the most popular technique for explaining scheduling mechanisms [32], which are obviously easier to understand and communicate than formulas full of mathematical symbols. An example is given in Fig. 3.10, in which the dynamic bottleneck detection (DBD) method proposed by Zhang et al. [33] is to be explained. Textual descriptions can be used to generate **global explanations** (for the scheduling system) or **local explanations** (for a specific job) such as.

- **Global explanation**: Based on the optimal schedule obtained using a genetic algorithm (GA), five of the twelve operations will be executed on machine #1.
- **Local explanation**: The first operation of job #1 is executed on machine #3 from time 0 to 4. Additionally, the second and third operations of the same job will be executed from time 4 to 8 and from time 8 to 12 on machines #1 and #3, respectively.

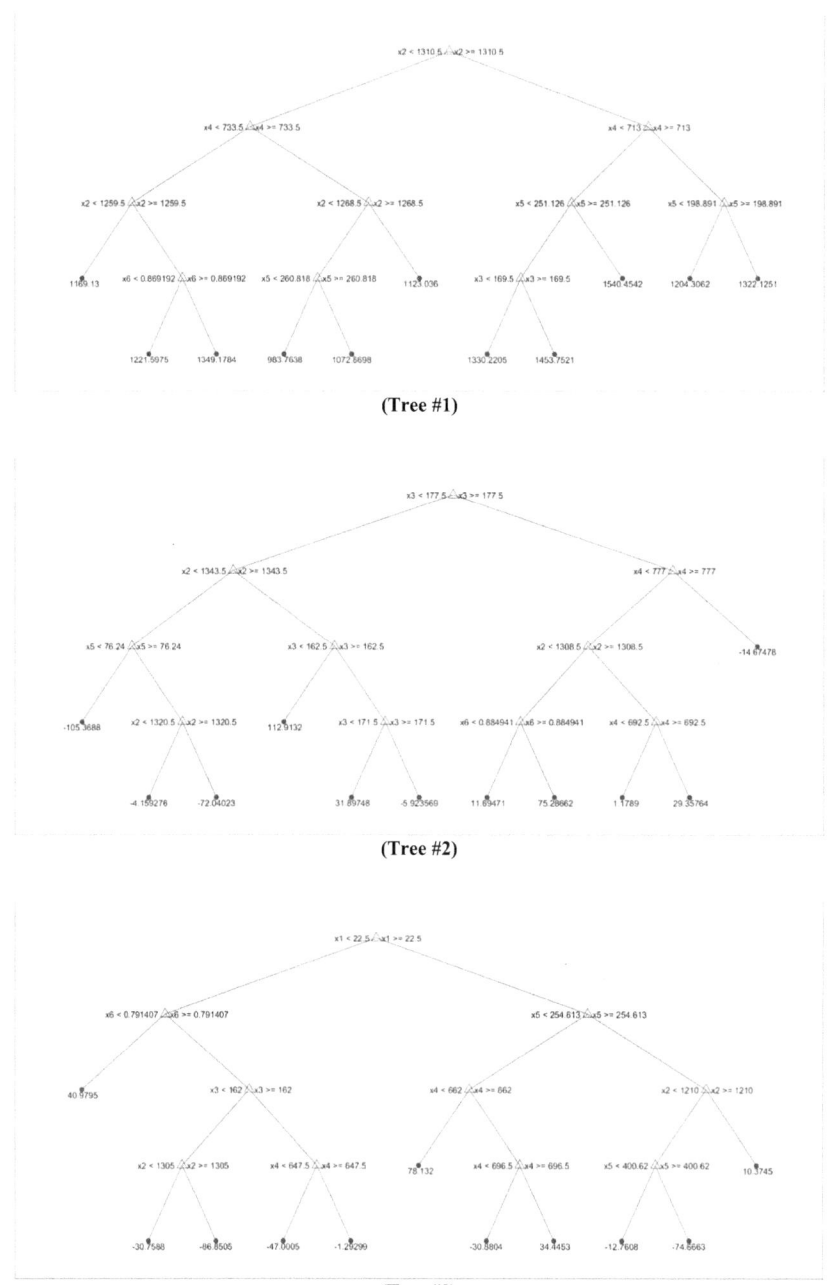

(Tree #1)

(Tree #2)

(Tree #3)

...

Fig. 3.9 GBDT for approximating the operations in the ANN for estimating the remaining cycle time

Table 3.1 Some dispatching rules for job sequencing and scheduling

Dispatching rule	Abbreviation	Job priority evaluation formula[*]
First in first out	FIFO	r_j
Critical ratio	CR	$\frac{d_j - t}{RPT_j}$
Cyclic scheduling	CYCLIC	(1) $\frac{1}{n_j \bmod k}$ (2) FIFO
Earliest due date	EDD	d_j
First in first out plus	FIFO+	(1) NQ_j + FIFO (2) FIFO
Fluctuation smoothing policy for mean cycle time	FSMCT	$j/\lambda - RCT_j$
Fluctuation smoothing policy for cycle time variation	FSVCT	$R_j - RCT_j$
Johnson's algorithm	–	(1) p_j ($\forall p_{j1} < p_{j2}$) (2) $1/p_j$ ($\forall p_{j1} \geq p_{j2}$) (3) (1) → (2)
Lawler's algorithm	–	(4) Plan backward (5)$f_j(C_j)$
Least amount of WIP in the next machine queue	LWNQ	NQ_j
Longest processing time until the next visit	LTNV	(1) $1/PTN_j$ (2) FIFO
Longest processing time	LPT	$1/p_j$
Shortest processing time	SPT	p_j
Shortest remaining processing time	SRPT	RPT_j
Shortest remaining processing time plus	SRPT+	(1) RPT_j (2) NQ_j
Shortest processing time until the next visit	STNV	(1) $1/PTN_j$ (2) FIFO
Weighted shortest processing time	WSPT	p_j/w_j

[*] Priority is the smaller the higher

Remark j: job no.; d_j: due date; k: cycle length; f_j: regular function; n_j: number of visits to the current machine; p_j: processing time; PTN_j: processing time until the next visit; r_j: arrival time; R_j: release time; RCT_j: remaining cycle time; RPT_j: remaining processing time; t: time; NQ_j: the queue length of the next machine; λ: release rate

Contrastive explanations can easily be made with textual descriptions, e.g.,

- **Local foil method**: When O_{23} (i.e., the third operation of job #2) precedes O_{32} on M4 (i.e., machine #4), then $C_{max} = 12$. In contrast, if O_{32} precedes O_{23} instead, C_{max} remains 12.

The following uses the dynamic bottleneck detection (DBD) method proposed by Zhang et al. [32] as an example. In DBD, different heuristics are used to sequence jobs before non-bottleneck and bottleneck workstations. First, jobs are divided into four categories. Then, jobs in these categories are sequenced using different dispatching rules [18]:

(1) First-priority category: CR is used to sequence jobs in this category first. Then, FIFO is applied to break possible ties (i.e., jobs with the same CRs).
(2) Second-priority category: The shortest processing time until the next bottleneck (SPNB) is applied to sequential jobs in this category. CR and FIFO are used in turn to break possible ties.
(3) Third-priority category: This category applies SPT, CR, and FIFO in turn.
(4) Fourth-priority category: CR and FIFO are used for this category.

Fig. 3.10 Text descriptions for explaining a scheduling mechanism

Another common textual technique for the same purpose is **pseudocode**, where the scheduling mechanism is explained informally and semi-colloquially in text similar to some programming languages [34] to facilitate system developers to write code that implements the scheduling mechanism, as illustrated in Fig. 3.11. Pseudocode is often used for providing **global explanations**.

Another way to evaluate the importance of an input is to observe its contribution to the job priority in the scheduling results. For example, FSVCT is applied to derive the job priority in a manufacturing system. To explain the scheduling mechanism of the FSVCT rule, a line chart is drawn to display and compare the slacks of all jobs, which is then compared with the scheduler's expectations of which jobs are most important (or urgent) and needed to be dealt with first (Fig. 3.12). In this way, a **local explanation** can be made for each job. In addition, by comparing the attributes (inputs) and slack (output) of all jobs, **contrastive explanations** are also possible. For example, in Fig. 3.13, the slack of job #11 is D rather than C, because its release time is B, not A, while the remaining cycle time estimate becomes irrelevant for the job.

The third textual technique for explaining a scheduling mechanism is to refer to its **scheduling performance**, such as the achieved makespan (C^*_{max}).

```
Begin
    Set r_min to a large value
    Set j_min to n+1 % null job
    For each i do
        If r_i is smaller than r_min
            update r_min to r_i
            update j_min to j_i
        End if
    End for
End
```

Fig. 3.11 Pseudocode for facilitating the implementation of the scheduling mechanism

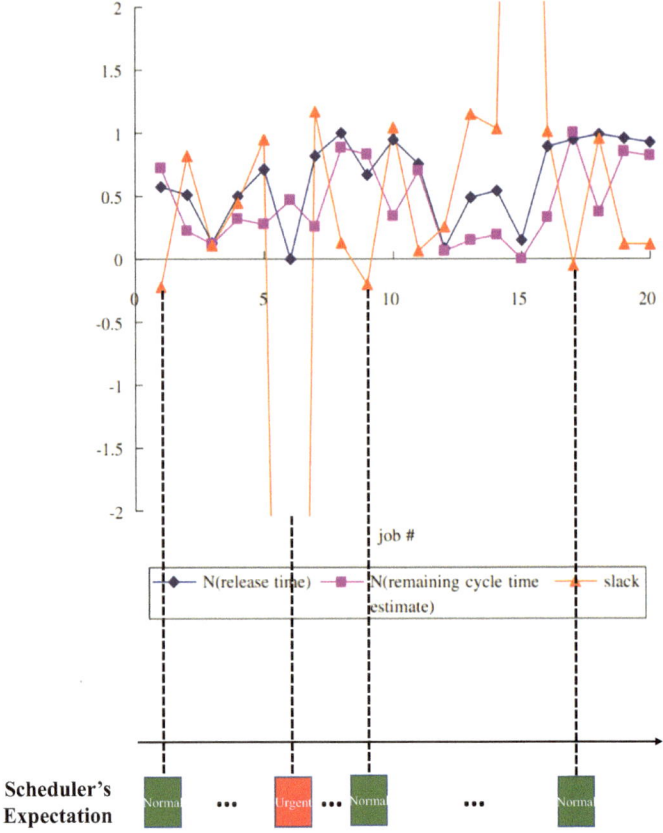

Fig. 3.12 Line chart for displaying and comparing the slacks of all jobs

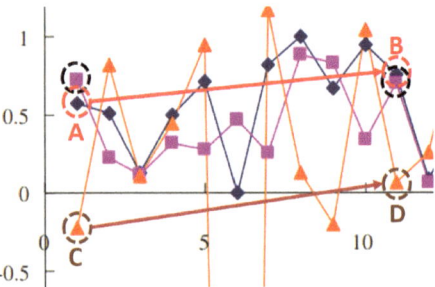

Fig. 3.13 Contrastive explanation

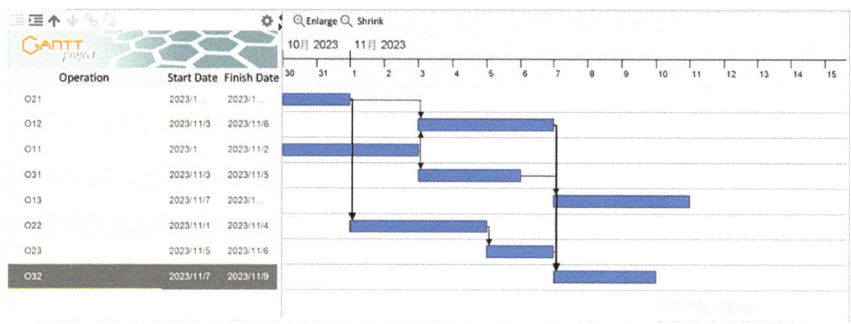

Fig. 3.14 Explaining a scheduling mechanism by visualizing the scheduling results using a Gantt chart

Gantt charts [34] are another common technique for explaining a scheduling mechanism by visualizing the scheduling results, as shown in Fig. 3.14.

Various **global**, **local**, **contrastive**, **if–then,** and **example-based explanations** can be generated based on the optimization results (Kamath and Liu, 2021), as summarized in Table 3.2.

Contrastive explanations can also be visualized using a Gantt chart, as shown in Fig. 3.15. In this diagram, the red operation is delayed by 2 days. As a result, the brown operation, as well as the makespan, will be delayed by 1 day.

Flowcharts have also been drawn to explain the scheduling mechanism in many past studies, as illustrated in Fig. 3.16.

The scheduling results generated by a scheduling system are good if.

- The schedule is feasible.
- The scheduling performance is efficient (ideally optimal).
- The scheduling results satisfy fixed (user) decisions.

After obtaining the optimal schedule, a bar chart can be used to compare the scheduling performances of various scheduling methods (see Fig. 3.17) [35].

Wang and Chen [36] modified the bar chart and proposed by applying common expressions, color management, and annotated figures [37]. The result is shown in Fig. 3.18, in which the following applies:

- All technical terms (i.e., symbols and variables) have been removed.
- The bar associated with the AI application is drawn in a different color (red).
- A horizontal line is drawn to mark the performance of the AI application as a benchmark.

Table 3.2 Global, local, and/or contrastive explanations based on the optimization results

Type of explanation	Examples
Global explanations	• Scheduling performance • Performances on other subsidiary indicators • Optimal schedule • Gantt chart illustrating the optimal schedule • Efficiency, execution time • Decision trees approximating the scheduling mechanism based on the optimization results • Etc.
Local explanations	• Production plan of a specific job: the start times, completion times, machines of all operations of a job; etc. • Sequence of all operations on a specific machine • Scheduling performance of a specific job: the completion time, tardy or not, tardiness, lateness, etc. • Scheduling performance of a specific machine: the number of operations processed, the average (maximum) number of waiting jobs, utilization, etc. • Etc.
Contrastive explanations (associated with minimal changes)	• Impact of swapping two consecutive operations (on a machine) on the scheduling performance • Impact of changing the machine of performing a specific operation on its completion time (or the scheduling performance) • Etc.
If-then explanations (nor necessarily associated with minimal changes)	• Sensitivity analysis • Parametric analysis • Etc.
Example-based explanations	• Characteristics of jobs performed on a specific machine • Categories of jobs with similar attributes and scheduling performance (such as cycle times, earliness, lateness, tardiness, etc.) • Etc.

3.3.2   SHAP Analysis

The **SHAP value** is defined as the weighted average of the marginal contributions of all possible coalitions $|\mathbf{F}|!$ as [38]:

$$\varphi_m(f) = \sum_{\{\mathbf{S} \subseteq \mathbf{F}\} \setminus \{m\}} \frac{|\mathbf{S}|!(|\mathbf{F}| - |\mathbf{S}| - 1)!}{|\mathbf{F}|!} \cdot [f(x_{\mathbf{S} \cup \{m\}}) - f(x_{\mathbf{S}})], \tag{3.11}$$

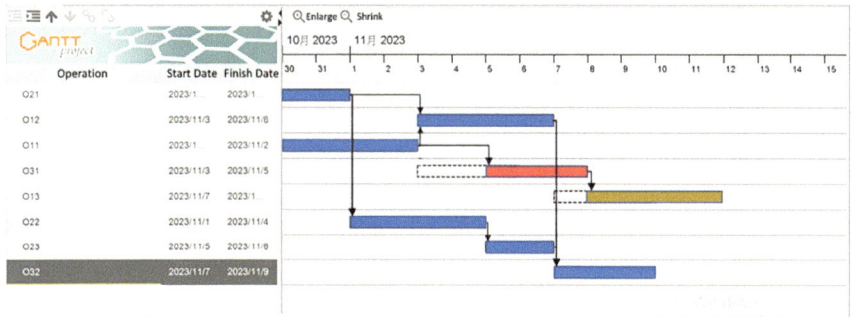

Fig. 3.15 Visualizing a contrastive explanation using a Gantt chart

Fig. 3.16 Flowchart for explaining STNV

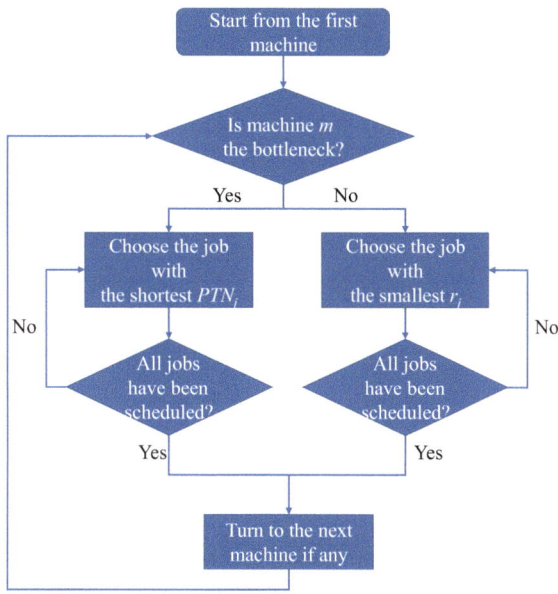

Fig. 3.17 Bar chart for comparing scheduling performances of various scheduling

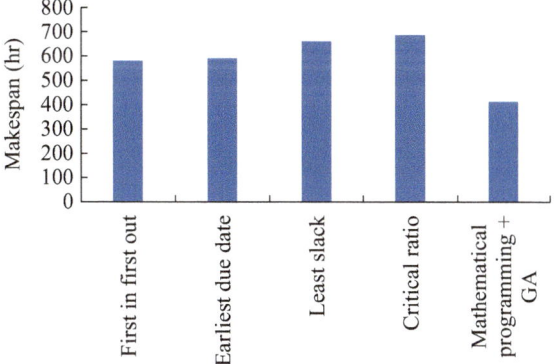

Fig. 3.18 Improved bar
chart

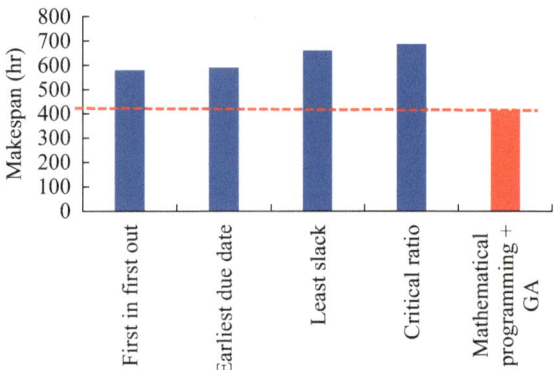

where $\varphi_m(f)$ is the weighted average Shapley value that feature m provides in the
context of all coalitions that exclude feature m. $\mathbf{F}$ is the set of all features; $\mathbf{S}$ is a
subset (i.e., coalition) of $\mathbf{F}$; $f(x_{\mathbf{S}\cup\{m\}})$ is the model prediction considering feature m,
while $f(x_\mathbf{S})$ is the model prediction without considering feature m.

SHAP evaluates the importance of an input by fixing its value while randomizing
other inputs [39]. Inputs to a scheduling problem include processing times, machine
capacities, machine available times, release times, due dates, job sizes, job priorities,
operation sequences, etc. Not all of these can be meaningfully randomized. For
example, the SHAP value of each operation can be evaluated as follows. First,

$$\mathbf{F} = \{(g_l,\ g_{l+1},\ g_{l+2})|g_l \in [1,\ n];\ g_{l+1} \in [1,\ K];\ g_{l+2} \in [1,\ m];$$
$$(g_{l_1},\ g_{l_1+1}) \neq (g_{l_2},\ g_{l_2+1})\ \forall l_1 \neq\ l_2;\ l,\ l_1,\ l_2 \in 3\cdot[1,\ nK]-2\} \qquad (3.12)$$

Assuming that the SHAP value of operation ξ is to be evaluated. Therefore, the
operation (including its position in the optimal solution and the machine on which it
is performed) is fixed:

$$(g_\xi, g_{\xi+1}, g_{\xi+2}) = (g_\xi^*, g_{\xi+1}^*, g_{\xi+2}^*) \qquad (3.13)$$

while the other operations are randomized:

$$(g_l, g_{l+1}, g_{l+2}) = (g_\zeta^*, g_{\zeta+1}^* U(1, K))\forall l \neq \xi; \zeta \neq \xi \qquad (3.14)$$

where U indicates uniform distribution. As a result,

$$\mathbf{S} = \{(g_l^*,\ g_{l+1}^*,\ g_{l+2}^*)\ |\ l \in 3[1,\ nK]-2\} \qquad (3.15)$$

$$\mathbf{S} \cup \{m\} = \{\{(g_\xi,\ g_{\xi+1},\ g_{\xi+2}), (g_l,\ g_{l+1},\ g_{l+2})\}\ |\ g_l \in [1,\ n];\ g_{l+1} \in [1,\ K];$$
$$\qquad (3.16)$$

Table 3.3 SHAP analysis results

Input	SHAP value
$X_{113} = 1$	-8
$X_{121} = 1$	-3.7
$X_{133} = 1$	-7
$X_{211} = 1$	-4.3
$X_{224} = 1$	-3
$X_{234} = 1$	-2.7
$X_{313} = 1$	-1.3
$X_{324} = 1$	-3.3

$$g_{l+2} \in [1, \ m]; \ (g_{l_1}, \ g_{l_1+1}) \neq (g_{l_2}, \ g_{l_2+1}) \ \forall l_1 \neq l_2;$$
$$l, \ l_1, \ l_2 \in 3 \cdot [1, \ nK] - 2 - \{\xi\}\} \tag{3.17}$$

In addition, f in Eq. (3.11) is equal to the scheduling performance.

A **SHAP analysis** can be performed in this way to evaluate the importance of each input to the output (i.e., the scheduling performance) for the optimal solution, for which the results of many iterations are averaged. The results are summarized in Table 3.3.

- The SHAP values of all inputs are negative, because the baseline is the optimal solution.
- The most important input to the optimal schedule is $X_{313} = 1$, i.e., O_{31} on M_3. By fixing this operation, the average fitness is just 1.3 below the optimal fitness. In other words, if the operation is not performed on M_3, the scheduling performance would be worse.

The results of a SHAP analysis are often summarized in a tornado plot (see Fig. 3.19), force plot (see Fig. 3.20), scatter plot (see Fig. 3.21), or beeswarm plot (see Fig. 3.22) [40].

3.3.3 Decision Tree-Based Methods

A decision tree is a special flowchart with many conditional judgments. Some scheduling rules are composite because they use multiple simple scheduling rules at the same time. For example, Johnson's algorithm [41] uses SPT and LPT in combination to schedule jobs with two operations: one that sorts jobs in which the first operation is shorter than the second and the other sorts jobs in which the first operation is longer than the second. The results of using the two simple scheduling rules are then combined to generate a complete schedule, which can be appropriately explained using a decision tree (see Fig. 3.23).

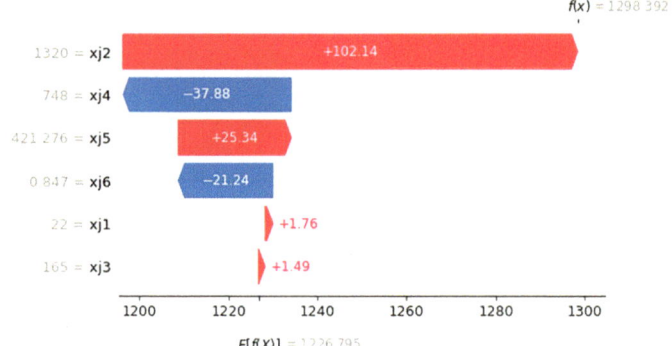

Fig. 3.19 Tornado plot for summarizing the results of a SHAP analysis (a single example)

Fig. 3.20 Force plot for the same purpose (a single example)

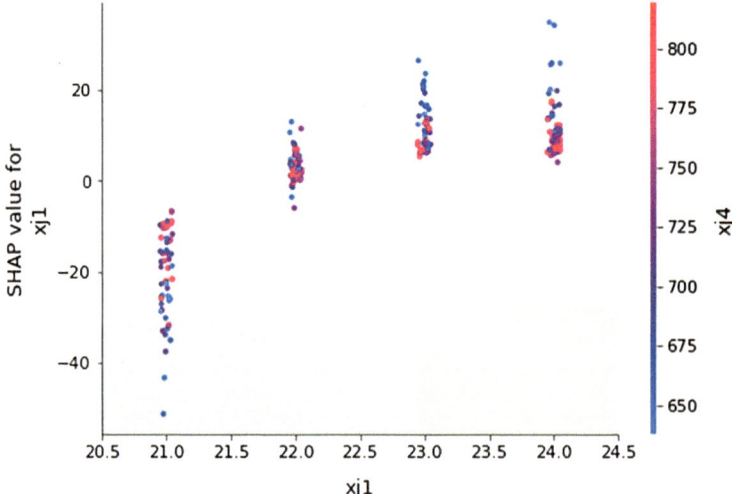

Fig. 3.21 Scatter plot for summarizing the results of a SHAP analysis (a single attribute; all examples)

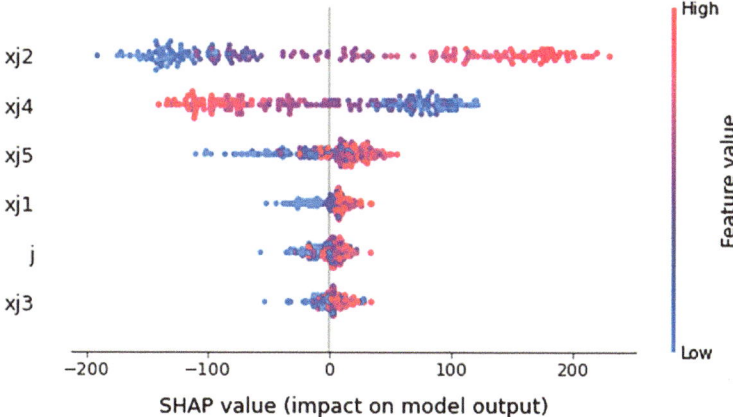

Fig. 3.22 Beeswarm plot for summarizing the results of a SHAP analysis (all attributes; all examples)

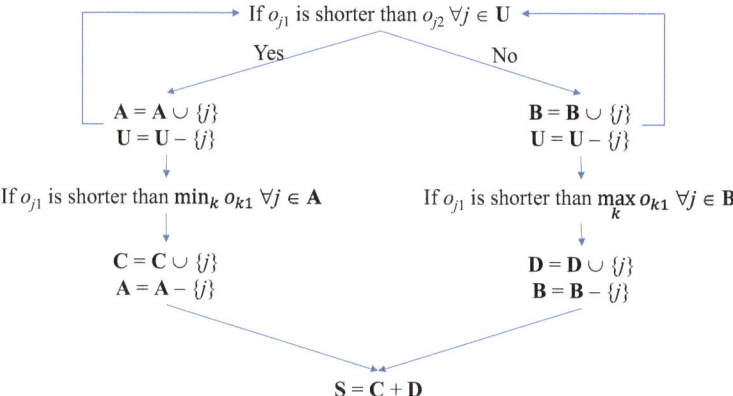

Fig. 3.23 Decision tree for explaining Johnson's algorithm

3.3.4   Lime

Some methods have been proposed in the literature to approximate the scheduling mechanism and results with a ANN. According to Weckman et al. [42], a trained ANN (not deep) can already successfully replicate the performances of genetic algorithms (GAs) on some benchmark problems. The ANN has the following configuration:

- $3nK$ Inputs ($\{(g_r, g_{r+1}, x_r(t))|r = 1 \sim nK\}$); $t = 1 \sim T$, corresponding to the job number, operation number, and encoded machine number of all operations. A machine number is encoded as

(Speed-Based Machine Encoding)

$$x_r(t) = \sum_{i \neq g_{r+2}} \varphi_{jkg_{r+2}i}(t) \tag{3.18}$$

$$\varphi_{jki_1 i_2}(t) = \begin{cases} 1 \text{ if} & p_{jki_1}(t) \leq p_{jki_2}(t) \\ 0 \text{ otherwise} \end{cases} \tag{3.19}$$

Other possible encoding methods include:

(i) Direct encoding:

$$x_r(t) = g_{r+2}. \tag{3.20}$$

(ii) Sequence codification schemes [43]:

(SCS 1)

$$x_r(t) = \{\text{BIN}(g_{r+2} - 1)\}, \tag{3.21}$$

where BIN is the binarification function.

(SCS 2)

$$x_r(t) = \{\alpha_q | \alpha_{g_{r+2}} = 1; \ \alpha_q = -1 \ \forall q \neq g_{r+2}; \ q = 1 \sim m\} \tag{3.22}$$

(SCS 3)

$$\{x_r(t)\} = \text{BIN}\left(\prod_{r=1}^{nK} g_{r+2}\right) \tag{3.23}$$

- No. of hidden layers: 1–2. The number of nodes in each hidden layer is enumerated and the one that provides the best training performance is selected [44].
- Output $o(t)$: the best scheduling performance that the scheduling mechanism can achieve.
- Approximation performance measure: root mean squared error (RMSE):

$$\text{RMSE} = \sqrt{\frac{\sum_{t=1}^{T} (o(t) - f(t))^2}{T}}. \tag{3.24}$$

Other performance measures are also applicable, such as mean absolute error (MAE), mean absolute percentage error (MAPE), and R^2:

$$\text{MAE} = \frac{\sum_{t=1}^{T} |o(t) - f(t)|^2}{T} \tag{3.25}$$

$$\text{MAPE} = \frac{\sum_{t=1}^{T} \frac{|o(t) - f(t)|}{f(t)}}{T} \cdot 100\% \tag{3.26}$$

$$R^2 = 1 - \frac{\sum_{t=1}^{T} (o(t) - f(t))^2}{\sum_{t=1}^{T} (\bar{f}(t) - f(t))^2} \tag{3.27}$$

- Training algorithm: Levenberg–Marquardt (LM) algorithm [45].
- Learning rate: 0.1–1.
- No. of epochs: 10,000–50,000.

In contrast, in Weckman et al. [42], the inputs to the ANN are the information (including the number, processing time, remaining processing time, machine load) of an operation, and the output is its position in the optimal schedule.

Subsequently, following the concept of local interpretable model-agnostic explanations (LIME) [46], a linear regression (LR) can be fitted to explain the scheduling mechanism locally, and then the effects of inputs can be assessed. The results are summarized in Fig. 3.24. In this figure, the makespan of the scheduling plan predicted using the LR is 11.76. In addition, the most important input is J8 (the job no. of the 8th operation in the scheduling plan), which is job #3, thereby increasing the makespan by 3.57 on average.

A RF can also be built to fit the trained ANN, so that simpler decision rules can be used to explain the scheduling process and results (see Fig. 3.25). To this end, synthetic data needs to be generated randomly yet logically based on the training data of ANN to consider the entire data space more comprehensively and reduce the impact of extreme cases. The trained ANN is then applied to predict the makespan of each schedule in the synthetic data.

There are L decision trees in RF, which are grown as follows. First, instead of all operations, each decision tree only considers randomly selected operations and

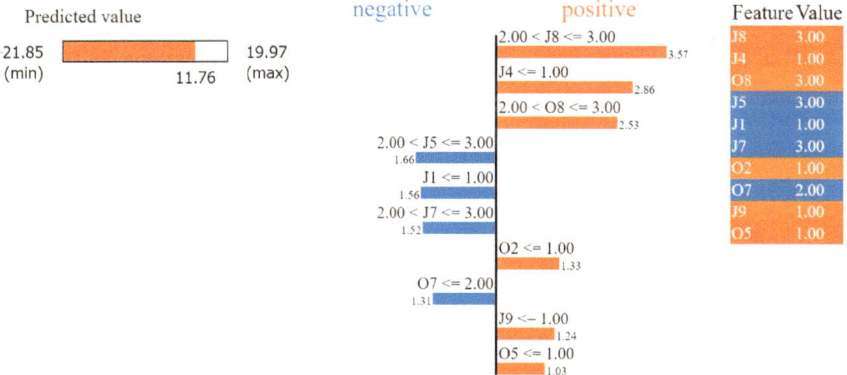

Fig. 3.24 Summary Report

Fig. 3.25 LIME application

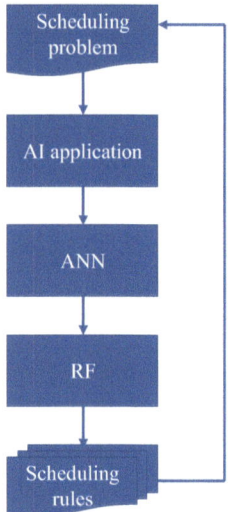

machines on which these operations are performed (i.e., feature bagging) [47]. Then, the schedules for decision tree fitting are also randomly selected (i.e., bootstrapped) from the training data [48]. The fitting results include the following decision rules.

$$\text{``If } x_1(t) \wedge_1 (\gamma, \ \tau)\xi_1(\gamma, \ \tau) \text{ and } \ldots \text{ and } x_{nK}(t) \wedge_{nK} (\gamma, \ \tau)\xi_{nK}(\gamma, \ \tau),$$
$$\text{then } \vartheta(t) = \beta(\gamma, \ \tau)\text{''}; l = 1 \sim L \tag{3.28}$$

where $\xi_r(\gamma, \ \tau)$ indicates the threshold for $x_r(t)$ in decision rule γ of decision tree τ; $\gamma = 1 \sim R$; $\tau = 1 \sim \psi$. Here, $\wedge_r(\gamma, \tau) \in \{=, <, \geq, \text{ within}\}$ is the logical operator applied; $\beta(\gamma, \ \tau)$ is the outcome of decision rule γ of decision tree τ that is used to approximate the DNN output as $\vartheta(t)$. For a schedule, the outcomes of the applicable decision rules in all decision trees are summarized as

$$\hat{f}(t) = \frac{\sum_{\tau=1}^{\Psi} \vartheta(t)}{\psi}$$
$$= \frac{\sum_{\tau=1}^{\Psi} \beta(\gamma, \tau)}{\psi}. \tag{3.29}$$

Then, the fitting accuracy can be measured as

$$\text{RMSE} = \sqrt{\frac{\sum_{t=1}^{T} (\hat{f}(t) - o(t))^2}{T}} \tag{3.30}$$

$$\text{MAE} = \frac{\sum_{t=1}^{T} |\hat{f}(t) - o(t)|^2}{T} \tag{3.31}$$

$$\text{MAPE} = \frac{\sum_{t=1}^{T} \frac{|\hat{f}(t) - o(t)|}{o(t)}}{T} \cdot 100\% \tag{3.32}$$

$$R^2 = 1 - \frac{\sum_{t=1}^{T} (\hat{f}(t) - o(t))^2}{\sum_{t=1}^{T} (\overline{o}(t) - o(t))^2}. \tag{3.33}$$

Decision rules like Eq. (3.28) are not only highly understandable, but can also be tailored to the manufacturing system. After converting the decision rules using common expressions [37], it is possible to generate explainable and customizable sequencing rules for any manufacturing system.

This study addresses this problem by using feasible solutions from all generations as inputs to the DNN. Even if these bionic computing algorithms start with the same initial solutions, their feasible solutions differ in subsequent generations due to their different evolutionary behaviors. The proposed methodology is therefore expected to provide different explanations for different bionic computing applications in job scheduling.

Example 3.1 A GA is applied to help solve the flexible job shop scheduling problem in [13]. Subsequently, a DNN is constructed to approximate the scheduling process and results using the GA with the following configuration:

- Twenty-four inputs, corresponding to the job number, operation number, and encoded machine number of all operations. The 2917 feasible yet non-repeating schedules of all generations in the GA before the convergence are used to train the DNN (see Table 3.4). The first 2800 examples are used as the training data, while the remaining is used for testing/evaluation.
- Two hidden layers with 48 and 6 neurons, respectively.
- Output: the approximate value of the makespan.
- Training algorithm: the LM algorithm with a learning rate of 0.1
- Convergence criteria: RMSE is less than 1 or 20,000 epochs have been run.

Table 3.4 Inputs and output of the DNN

t	g_1	g_2	g_3	g_4	g_5	g_6	...	g_{22}	g_{23}	g_{24}	C_{max}
1	1	1	4	2	1	2		3	2	3	35
2	3	1	4	1	1	3		3	3	1	35
3	2	1	2	3	1	3		1	3	1	32
4	3	1	2	1	1	4		2	2	4	29
5	1	1	2	2	1	2		3	2	1	18
6	3	1	1	3	2	4		1	3	4	29
...											
2917	2	1	2	3	1	4		2	2	2	23

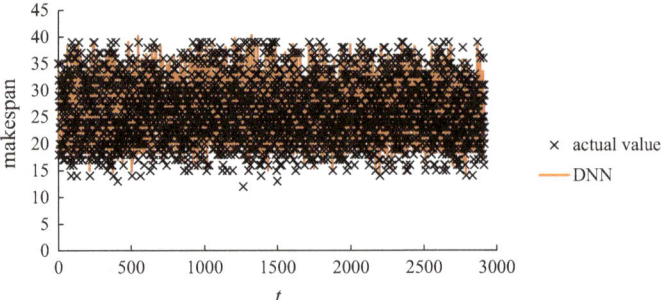

Fig. 3.26 Approximation results

The approximation results are shown in Fig. 3.26. The approximation performance for test data is evaluated as

$$MAE = 1.12$$
$$MAPE = 4.7\%$$
$$RMSE = 1.68$$
$$R^2 = 0.96$$

showing a promising approximation accuracy.

Example 3.2 A RF with 10 decision trees is built to fit the trained DNN in Example 3.1 to generate explainable and customized scheduling rules. To this end, synthetic data is generated based on the training data of DNN to consider the entire data space more comprehensively and reduce the impact of extreme cases. The trained DNN is then applied to predict the makespan of each example in the synthetic data. The built RF is shown in Fig. 3.27. The decision rules of these decision trees are summarized in Table 3.5. The fitting accuracy is evaluated as

$$MAE = 0.87$$
$$MAPE = 3.6\%$$
$$RMSE = 1.19$$
$$R^2 = 0.95$$

Taking the third feasible schedule {(2, 1, 2), (3, 1, 3), (2, 2, 1), (2, 3, 3), (1, 1, 3), (1, 2, 4), (3, 2, 3), (1, 3, 1)} as an example. In each decision tree, there is a decision rule that is applicable to the schedule, as summarized in Table 3.6. After aggregation, the makespan of the schedule is fitted as 29.1. The actual value is 32, while the approximate value obtained using DNN is 31.2. The decision rules are then converted using common expressions into scheduling rules that are presented in Table 3.7.

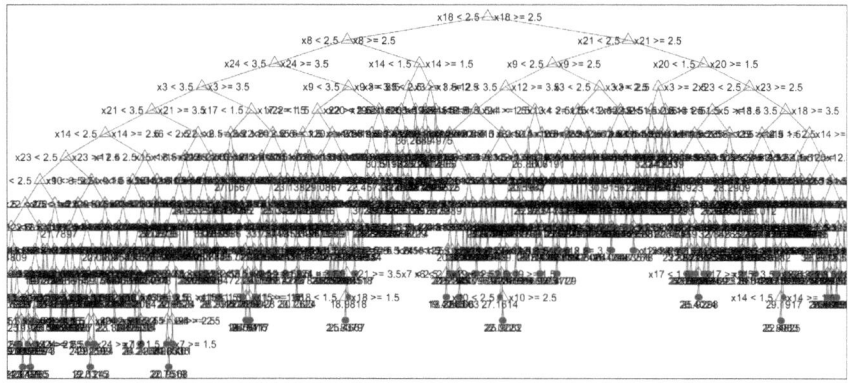

(Decision tree #1)

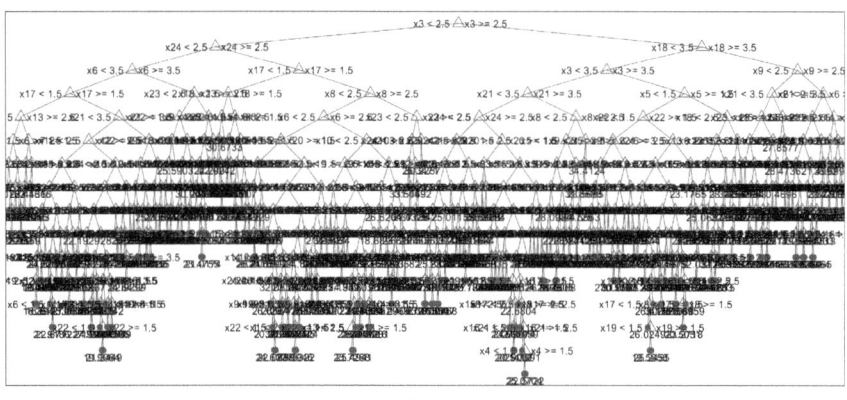

(Decision tree #2)

...

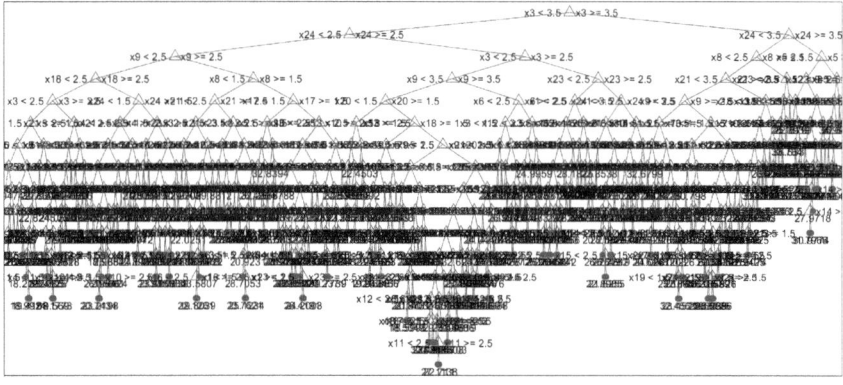

(Decision tree #10)

Fig. 3.27   Built RF for the case

Table 3.5 Decision rules of the RF (in MATLAB format)

Decision tree #	Decision rules
1	1 if $\times$ 18 < 2.5 then node 2 elseif $\times$ 18 > = 2.5 then node 3 else 25.9447 2 if $\times$ 8 < 2.5 then node 4 elseif $\times$ 8 > = 2.5 then node 5 else 25.1568 3 if $\times$ 21 < 2.5 then node 6 elseif $\times$ 21 > = 2.5 then node 7 else 26.7776 … 895 fit = 22.1243 896 fit = 22.7519 897 fit = 20.9568
2	1 if $\times$ 3 < 2.5 then node 2 elseif $\times$ 3 > = 2.5 then node 3 else 26.141 2 if $\times$ 24 < 2.5 then node 4 elseif $\times$ 24 > = 2.5 then node 5 else 25.2807 3 if $\times$ 18 < 3.5 then node 6 elseif $\times$ 18 > = 3.5 then node 7 else 27.0137 … 905 fit = 25.2455 906 fit = 25.0704 907 fit = 22.5722
…	
10	1 if $\times$ 3 < 3.5 then node 2 elseif $\times$ 3 > = 3.5 then node 3 else 26.0547 2 if $\times$ 24 < 2.5 then node 4 elseif $\times$ 24 > = 2.5 then node 5 else 25.4384 3 if $\times$ 24 < 3.5 then node 6 elseif $\times$ 24 > = 3.5 then node 7 else 28.0104 … 903 fit = 22.4703 904 fit = 22.711 905 fit = 27.1138

Table 3.6 Decision rules applicable to the first feasible schedule

Decision rule #	Rule content
1	If $2.5 <\ =\ \times 12 < 3.5$ And $\times 14 < 1.5$ And $\times 15 < 3.5$ And $\times 18 > = 3.5$ And $\times 20 > = 1.5$ And $\times 21 > = 2.5$ And $\times 23 > = 2.5$ Then $C_{max} = 31.6654$
2	If $\times 3 > = 1.5$ And $\times 6 < 3.5$ And $\times 8 < 2.5$ And $\times 12 > = 2.5$ And $\times 14 < 1.5$ And $\times 17 > = 1.5$ And $\times 18 > = 1.5$ And $\times 21 < 3.5$ And $\times 22 < 2.5$ And $\times 24 < 2.5$ Then $C_{max} = 29.8027$
…	
10	If $\times 1 > = 1.5$ And $\times 4 > = 2.5$ And $\times 7 > = 1.5$ And $\times 9 < 1.5$ And $\times 13 < 2.5$ And $\times 18 > = 2.5$ And $\times 24 < 1.5$ Then $C_{max} = 26.6072$

According to the analysis results, the following discussion is made:

- The approximation accuracy using the DNN is satisfactory since the RMSE is less than 2, which indicated that the fitness (or makespan) of every feasible solution (or schedule) generated during the evolution process of the GA can be estimated using the trained DNN.
- Although the approximation accuracy using speed-based machine encoding is better than the other four existing encoding mechanisms, the advantage is not very significant (see Table 3.8). However, these encoding mechanisms change the

Table 3.7 Scheduling rules

Decision rule no	Decision rule content
1	If the speed of the machine of the 4th operation is the 3rd And the 1st operation of the 5th job is performed And the speed of the machine of the 5th operation is faster than the 4th And the speed of the machine of the 6th operation is the slowest (4th) And the 2nd or 3rd operation of the 7th job is performed on a machine faster than the 2nd And the 3rd operation of the 8th job is performed Then C_{max} is approximately 31.6654
2	If the speed of the machine of the 1st operation is slower than the 1st And the speed of the machine of the 2nd operation is faster than the 4th And the 1st or 2nd operation of the 3rd job is performed And the speed of the machine of the 4th operation is slower than the 2nd And the 1st operation of the 5th job is performed And the 2nd, 3rd or 4th operation of the 6th job is performed And the speed of the machine of the 6th operation is slower than the 1st And the speed of the machine of the 7th operation is faster than the 4th And the 8th job is J1 or J2 And the speed of the machine of the 8th operation is faster than the 3rd Then $C_{max} = 29.8027$
...	
10	If the 1st job is J2 or J3 And the 2nd job is J3 And the 3rd job is J2 or J3 And the speed of the machine of the 3rd operation is the fastest (1st) And the 5th job is J1 or J2 And the speed of the machine of the 6th operation is slower than the 2nd And the speed of the machine of the 8th operation is the fastest (1st) Then $C_{max} = 26.6072$

formats of the decision rules, which affects the ease of interpreting the scheduling rules. Speed-based machine encoding differentiates machines by their speeds. In this way, larger machine numbers corresponded to slower machines, which is more logical.

Table 3.8 Approximation performances of various encoding mechanisms

Encoding mechanism	MAE	MAPE (%)	RMSE	R^2
Direct encoding	1.55	6.2	2.00	0.87
SCS1	1.58	6.4	2.09	0.86
SCS2	1.57	6.3	2.00	0.87
SCS3	1.57	6.3	2.00	0.87
Speed-based machine encoding	1.12	4.7	1.68	0.96

- In addition, applying the trained DNN is more efficient than running the time-consuming GA.
- The choice of the encoding method affects the content of the generated scheduling rules and also affects the ease of explaining the scheduling rules.
- The scheduling rules generated by the RF tell the scheduling personnel the results of making specific arrangements in the scheduling plan. The scheduling personnel can then avoid making arrangements that lead to poor scheduling performance and make those that might shorten the makespan. For example, in Table 3.7, the outcome of the 10th decision rule shows that it can lead to a shorter makespan, so the scheduling personnel can arrange jobs, processes, and machines based on this decision rule. For example, J2 (O21) or J3 (O31) should be the first job (or operation) assigned, but not on which machine. In contrast, the outcome of the first decision rule results in a longer makespan. Therefore, the scheduling personnel should not make arrangements similar to the conditions of the decision rule.
- The same GA can be applied to solve job scheduling problems in different manufacturing systems. In contrast, the decision rules of the RF generated for different manufacturing systems are not the same, which enables the generation of customizable scheduling rules tailored to specific manufacturing systems.

The decision trees constructed in Fig. 3.27 estimate the scheduling performance based on job sequences, which may not be very straightforward (or useful) for guiding scheduling operations. To enhance the effectiveness of GenAI, the inputs and outputs of the DNN should be better selected, as shown in Fig. 3.28.

Most model-agnostic post hoc XAI methods work on relationships between the inputs and output without reference to the architecture of the AI model. However, bionic computing algorithms such as GA, ABC, and ACO may have the same inputs and output. Distinguishing these bionic computing algorithms becomes difficult using model-agnostic post hoc XAI methods. This problem can be addressed by using feasible solutions from all generations as inputs to the DNN. Even if these

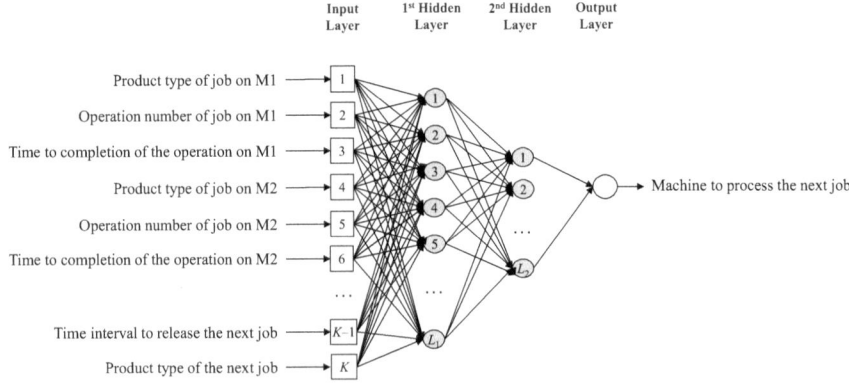

Fig. 3.28 Better selection of the inputs and output of the DNN

bionic computing algorithms start with the same initial solutions, their feasible solutions differ in subsequent generations due to their different evolutionary behaviors, thereby providing different explanations for different bionic computing applications in job scheduling.

3.3.5   Explainable Optimization

Job scheduling problems are often formulated as mathematical programming (MP) (i.e., optimization) problems. Although the objective function and constraints reflect the reality of the manufacturing system and are easily understood by factory workers, the optimization solvers used to solve the MP problem are considered black boxes [49], even though AI technologies are not applied therein and only traditional solution techniques like branch and bound [50]. To solve this problem, explainable optimization needs to be pursued instead of XAI. In the opinion of Čyras et al. [49], argumentation, a technique for explaining reasoning and decision-making, is applicable [50]. Argumentative explanations have been provided in various settings, and abstract arguments (AA) can present the accessible knowledge of an optimizer. To this end, they proposed AA frameworks (AFs) for modeling optimization problems, i.e., ArgOpt.

3.4   Criteria for Evaluating the Effectiveness of XAI Applications in Job Sequencing and Scheduling

In the view of Čyras et al. [49], a good explanation should be efficiently achievable, incorporate few causal connections, and accept simple natural language explanations. In addition,

- To build trust, explanations should be linked to formal representations, providing explainable credentials as to why the explanation is valid and how it was generated.
- For tractability, polynomial interpretability can be pursued for computational tractability (i.e., fast generation of results) and cognitive tractability (i.e., clear user interpretation). Using direct polynomials to approximate iterative optimization processes or calculations involving special mathematical functions (e.g., step, exponential, logarithmic, and triangular functions) does facilitate rapid understanding by the user [51–53]. However, in Čyras et al. [49], computational tractability means that it is possible to explain whether and why a schedule is "good" in polynomial time, while cognitive tractability means that the size of each explanation presented to the user should be a polynomial function of the size of the scheduling problem.
- Various types of explanations can be generated.
- The explanation mechanism adopts a modular structure.

Fig. 3.29 Requirements for effective XAI applications in job sequencing and scheduling

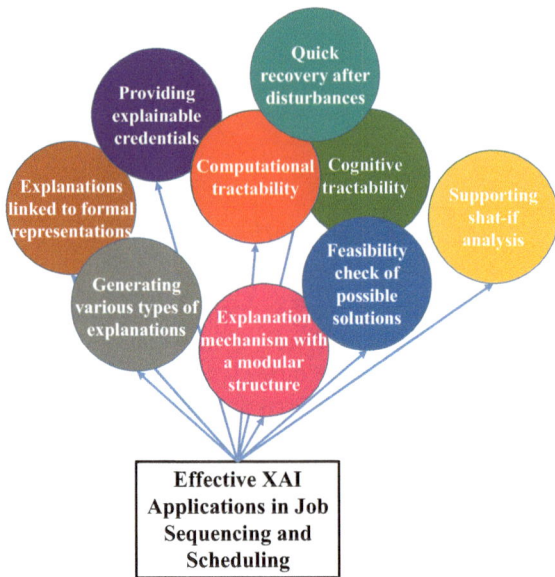

- The explanation mechanism enables the decision-maker to check the feasibility of possible solutions, conduct what-if analysis for scenarios, and recover feasible solutions after various disturbances as summarized in Fig. 3.29.

According to Chen [32], XAI techniques for interpreting AI applications in job sequencing and scheduling are considered effective if the following requirements are met:

- Personnel responsible for job scheduling have the required background knowledge.
- A XAI technique can process high-dimensional scheduling data.
- A XAI technique is convenient for comparing the performances of various scheduling methods.
- Explanation formats are consistent in different applications.
- A XAI technique is easy to communicate.
- A XAI technique is simple and easy to understand.
- A XAI technique can visualize/compare feasible solutions.
- A XAI technique can visualize/track the evolution process.

Various XAI techniques for explaining AI applications in job scheduling are compared in Table 3.9 in terms of effectiveness. Most existing XAI techniques can only meet up to five requirements. There is no perfect XAI technique to explain the scheduling process and results. Furthermore, existing XAI techniques can be improved, and new (or tailored) ones can be proposed. Although each has its advantages, utilizing multiple XAI techniques at the same time can compensate for the shortcomings of each other and offer a comprehensive solution to explain scheduling operations.

Table 3.9 Comparison of existing XAI techniques

	I	II	III	IV	V	VI	VII	VIII
Requires background knowledge	V	V		V	V	V	V	V
Can handle high-dimensional data				V				V
Compares scheduling performances						V	V	
Has consistent explanation format				V	V	V	V	V
Is easy to communicate	V	V		V	V	V	V	V
Is easy to understand	V	V		V	V	V	V	V
Explains crossover operation	V	V	V	V				V
Explains mutation operation	V	V	V	V				V
Explains selection operation	V	V	V	V				V
Visualizes contribution of feasible solutions								
Visualizes/compares feasible solutions				V				
Visualizes/traces evolution process				V	V			

I: Textual description; II: Flowchart; III: Pseudocode; IV: Chromosome diagram; V: Dynamic line chart; VI: Bar chart; VII: Bar chart; VIII: Decision tree-based interpretation

References

1. T. Chen, Y.C. Wang, A nonlinear scheduling rule incorporating fuzzy-neural remaining cycle time estimator for scheduling a semiconductor manufacturing factory—a simulation study. Int. J. Adv. Manuf. Technol. **45**, 110–121 (2009)
2. I. Ullah, M.S. Khan, M. Amir, J. Kim, S.M. Kim, LSTPD: Least slack time-based preemptive deadline constraint scheduler for Hadoop clusters. IEEE Access **8**, 111751–111762 (2020)
3. T.C. Chiang, L.C. Fu, Solving the FMS scheduling problem by critical ratio-based heuristics and the genetic algorithm. Proc. IEEE Int. Conf. Robot. Autom. **3**, 3131–3136 (2004)
4. J.R. Morrison, B. Campbell, E. Dews, J. LaFreniere, Implementation of a fluctuation smoothing production control policy in IBM's 200mm wafer fab, in *Proceedings of the 44th IEEE Conference on Decision and Control* (2005), pp. 7732–7737
5. T. Chen, A hybrid look-ahead SOM-FBPN and FIR system for wafer-lot-output time prediction and achievability evaluation. Int. J. Adv. Manuf. Technol. **35**, 575–586 (2007)
6. P. Melin, J.C. Monica, D. Sanchez, O. Castillo, Analysis of spatial spread relationships of coronavirus (COVID-19) pandemic in the world using self organizing maps. Chaos Solitons Fractals **138**, 109917 (2020)
7. T. Chen, Y.-C. Wang, H.-R. Tsai, Lot cycle time prediction in a ramping-up semiconductor manufacturing factory with a SOM-FBPN-ensemble approach with multiple buckets and partial normalization. Int. J. Adv. Manuf. Technol. **42**(11–12), 1206–1216 (2009)
8. A. Flexer, On the use of self-organizing maps for clustering and visualization. Intelligent Data Analysis **5**(5), 373–384 (2001)
9. J. Dombi, O. Csiszár, *Explainable Neural Networks Based on Fuzzy Logic and Multi-Criteria Decision Tools* (Springer, 2021)
10. T. Chen, C.-W. Lin, Y.-C. Lin, A fuzzy collaborative forecasting approach based on XAI applications for cycle time range estimation. Appl. Soft Comput. **151**, 111122 (2024)
11. A. Fernandez, F. Herrera, O. Cordon, M.J. del Jesus, F. Marcelloni, Evolutionary fuzzy systems for explainable artificial intelligence: why, when, what for, and where to? IEEE Comput. Intell. Mag. **14**(1), 69–81 (2019)

12. T.C.T. Chen, Job sequencing and scheduling, in *Production Planning and Control in Semiconductor Manufacturing: Big Data Analytics and Industry 4.0 Applications* (2020), pp. 77–99.
13. F. Pezzella, G. Morganti, G. Ciaschetti, A genetic algorithm for the flexible job-shop scheduling problem. Comput. Oper. Res. **35**(10), 3202–3212 (2008)
14. P. Sedgwick, Pearson's correlation coefficient. Bmj 345 (2012)
15. P. Sedgwick, Spearman's rank correlation coefficient. Bmj 349 (2014)
16. M. Green, U. Ekelund, L. Edenbrandt, J. Björk, J.L. Forberg, M. Ohlsson, Exploring new possibilities for case-based explanation of artificial neural network ensembles. Neural Netw. **22**(1), 75–81 (2009)
17. C. Molnar, 9.5 Shapley values (2022). https://christophm.github.io/interpretable-ml-book/sha pley.html#shapley
18. H. Hamilton, Cumulative gains and lift charts (2009). https://www2.cs.uregina.ca/~dbd/cs831/notes/lift_chart/lift_chart.html
19. B. Keating, Lift charts (2024). https://www3.nd.edu/~busiforc/handouts/DataMining/Lift%20Charts.html
20. A., Shrikumar, P., Greenside, A. Kundaje, Learning important features through propagating activation differences, in *International Conference on Machine Learning* (2017), pp. 3145–3153
21. H. Chen, S.M. Lundberg, S.I. Lee, Explaining a series of models by propagating Shapley values. Nat. Commun. **13**(1), 4512 (2022)
22. P. Schwab, W. Karlen, Cxplain: causal explanations for model interpretation under uncertainty, in *Advances in Neural Information Processing Systems* (2019), pp. 1–11
23. T. Chen, Y.-C. Lin, Fuzzified deep neural network ensemble approach for estimating the cycle time range. Appl. Soft Comput. **130**, 109697 (2022)
24. B. Waschneck, A. Reichstaller, L. Belzner, T. Altenmüller, T. Bauernhansl, A. Knapp, A. Kyek, Deep reinforcement learning for semiconductor production scheduling, in *2018 29th Annual SEMI Advanced Semiconductor Manufacturing Conference* (2018), pp. 301–306
25. E.M. Kenny, M.T. Keane, Twin-systems to explain artificial neural networks using case-based reasoning: comparative tests of feature-weighting methods in ANN-CBR twins for XAI, in *Twenty-Eighth International Joint Conferences on Artificial Intelligence* (2019), pp. 2708–2715
26. A. Likas, N. Vlassis, J.J. Verbeek, The global k-means clustering algorithm. Pattern Recogn. **36**(2), 451–461 (2003)
27. T. Chen, An intelligent hybrid system for wafer lot output time prediction. Adv. Eng. Inform. **21**, 55–65 (2007)
28. W.Y. Loh, Classification and regression trees. Wiley Interdiscip. Rev.: Data Min. Knowl. Discov. **1**(1), 14–23 (2011)
29. J. Liu, Q. Huang, C. Ulishney, C.E. Dumitrescu, Comparison of random forest and neural network in modeling the performance and emissions of a natural gas spark ignition engine. J. Energy Res. Technol. **144**(3), 032310 (2022)
30. GoogleDevelopers, Gradient boosted decision trees | Machine learning (2022). https://developers.google.com/machine-learning/decision-forests/intro-to-gbdt
31. T. Chen, T. He, M. Benesty, V. Khotilovich, Y. Tang, H. Cho, K. Chen, Xgboost: Extreme gradient boosting. R Package Version 0.4–2, **1**(4), 1–4 (2015)
32. T.C.T. Chen, Applications of XAI to job sequencing and scheduling in manufacturing, in *Explainable Artificial Intelligence (XAI) in Manufacturing: Methodology, Tools, and Applications* (2023), pp. 83–105
33. H. Zhang, Z. Jiang, C. Guo, Simulation-based optimization of dispatching rules for semiconductor wafer fabrication system scheduling by the response surface methodology. Int. J. Adv. Manuf. Technol. **41**(1–2), 110–121 (2009)
34. H.Z. Jia, J.Y. Fuh, A.Y. Nee, Y.F. Zhang, Integration of genetic algorithm and Gantt chart for job shop scheduling in distributed manufacturing systems. Comput. Ind. Eng. **53**(2), 313–320 (2007)

35. T. Chen, A fuzzy-neural DBD approach for job scheduling in a wafer fabrication factory. Int. J. Innov. Comput. Inf. Control. **8**(6), 4024–4044 (2012)

36. Y.C. Wang, T. Chen, Adapted techniques of explainable artificial intelligence for explaining genetic algorithms on the example of job scheduling. Expert Syst. Appl. **237**, 121369 (2024)

37. Y.C. Lin, T.C.T. Chen, Type-II fuzzy approach with explainable artificial intelligence for nature-based leisure travel destination selection amid the COVID-19 pandemic. Digital Health **8**, 20552076221106320 (2022)

38. L.P. Joseph, E.A. Joseph, R. Prasad, Explainable diabetes classification using hybrid Bayesian-optimized TabNet architecture. Comput. Biol. Med. **151**, 106178 (2022)

39. Y. Meng, N. Yang, Z. Qian, G. Zhang, What makes an online review more helpful: an interpretation framework using XGBoost and SHAP values. J. Theor. Appl. Electron. Commer. Res. **16**(3), 466–490 (2020)

40. GitHub, shap/shap (2024). https://github.com/shap/shap

41. H. Allaoui, A. Artiba, Johnson's algorithm: A key to solve optimally or approximately flow shop scheduling problems with unavailability periods. Int. J. Prod. Econ. **121**(1), 81–87 (2009)

42. G.R. Weckman, C.V. Ganduri, D.A. Koonce, A neural network job-shop scheduler. J. Intell. Manuf. **19**, 191–201 (2008)

43. D.J. Fonseca, D. Navaresse, Artificial neural networks for job shop simulation. Adv. Eng. Inform. **16**(4), 241–246 (2002)

44. D. Stathakis, How many hidden layers and nodes? Int. J. Remote Sens. **30**(8), 2133–2147 (2009)

45. J. Nocedal, S. Wright, *Numerical Optimization* (Springer Science & Business Media, 2006)

46. M.R. Zafar, N. Khan, Deterministic local interpretable model-agnostic explanations for stable explainability. Mach. Learn. Knowl. Extr. **3**(3), 525–541 (2021)

47. T.C.T. Chen, Applications of XAI for forecasting in the manufacturing domain, in *Explainable Artificial Intelligence (XAI) in Manufacturing: Methodology, Tools, and Applications* (2023), pp. 13–50

48. V. Rodriguez-Galiano, M. Sanchez-Castillo, M. Chica-Olmo, M.J.O.G.R. Chica-Rivas, Machine learning predictive models for mineral prospectivity: an evaluation of neural networks, random forest, regression trees and support vector machines. Ore Geol. Rev. **71**, 804–818 (2015)

49. K. Čyras, D. Letsios, R. Misener, F. Toni, Argumentation for explainable scheduling. Proc. AAAI Conf. Artif. Intell. **33**(01), 2752–2759 (2019)

50. A. D'ariano, D. Pacciarelli, M. Pranzo, A branch and bound algorithm for scheduling trains in a railway network. Eur. J. Oper. Res. **183**(2), 643–657 (2007)

51. K. Atkinson, P. Baroni, M. Giacomin, A. Hunter, H. Prakken, C. Reed, G.R. Simari, M. Thimm, S. Villata, Towards artificial argumentation. AI Mag. **38**(3), 25–36 (2017)

52. T. Chen, H.-C. Wu, M.-C. Chiu, A deep neural network with modified random forest incremental interpretation approach for diagnosing diabetes in smart healthcare. Appl. Soft Comput. **152**, 111183 (2024)

53. T. Chen, Y.-C. Wang, A modified random forest incremental interpretation method for explaining artificial and deep neural networks in cycle time prediction. Decis. Anal. **7**, 100226 (2023)

Chapter 4
Explaining Genetic Algorithm Applications in Job Sequencing and Scheduling

4.1 GA Applications in Job Sequencing and Scheduling

Genetic algorithms (GA) have been widely used in job sequencing and scheduling [1–5]. GA helps explain the process of solving complex job sequencing and scheduling problems by describing how to improve the feasible solution [6]. In other words, GA describes the evolution of feasible solutions to optimal solutions for mathematical programming problems of job sequencing and scheduling [7].

Applications of GA are of particular interest because such applications are most common in job sequencing and scheduling [8]. Some recent examples are given below.

Davis [9] applied a GA to deal with non-deterministic problems using multiple abstraction levels and progressive constraint relaxation in a framework-based representation system to explain the scheduling results of a job shop scheduling system.

Salido et al. [10] proposed a GA to solve a job shop scheduling problem, where machines consumed different energy to process tasks at different rates (speed scaling). They did not formulate the shop scheduling problem as a mathematical programming (MP) problem. As long as a special fitness function that takes energy consumption into account is defined, such a job shop scheduling problem can be solved directly using GAs.

Qing-dao-er-ji and Wang [11] proposed a new hybrid GA to solve a job shop scheduling problem, in which a hybrid selection operator based on the fitness value and the concentration value was used. Additionally, a new crossover operator was defined per machine. The mutation operator took into account operations on the critical path. They also employed a local search mechanism to improve the capability of the GA.

De Giovanni and Pezzella [12] considered a distributed flexible job shop scheduling problem (DFJS) consisting of multiple flexible manufacturing units (FMUs). The goal was to minimize the global completion time for all FMUs. They

© The Author(s), under exclusive license to Springer Nature Switzerland AG 2025
T. T. Chen, *Explainable and Customizable Job Sequencing and Scheduling*,
SpringerBriefs in Applied Sciences and Technology,
https://doi.org/10.1007/978-3-031-85374-6_4

proposed a GA to solve the DFJS problem, where the genetic encoding contained information about allocating jobs to FMUs. Furthermore, the greedy decoding process exploited flexibility and determined job routing. Similar to Qing-dao-er-ji and Wang [11], they also adopted a local search mechanism to improve the available solutions by refining the most promising solutions of each generation.

However, many of the GA applications in past research have not been thoroughly explained [13]. For example, Pezzella et al. [14] solved the four-machine flexible job shop scheduling problem aiming to minimize the makespan, namely a $FJ4//C_{max}$ scheduling problem, in which different jobs on the same machine have different processing times. Furthermore, each operation can be performed on different machines with unequal processing times. Theoretically, the scheduling problem can be formulated as a mixed integer-nonlinear programming (MINLP) problem (see Fig. 4.1), which is NP-hard. The number of possible permutations for this problem is at most $4^{3+3+2} \cdot 8! = 2.64 \cdot 10^9$. Pezzella et al. applied a GA to help solve the MINLP problem. In fact, there is no need to formulate the MINLP problem before applying the GA. Feasible schedules (in permutation representation) can be easily generated for this problem. A simplified production simulator can also be applied to derive the makespan associated with each feasible schedule.

4.2 Explaining GA Applications in Job Sequencing and Scheduling—Traditional Approaches

Chromosome diagrams (or maps) are often used to illustrate chromosome coding (i.e., permutation representation) and operations (e.g., **crossover** and **mutation**) [8, 15, 16], in which solutions to a job sequencing and scheduling problem are represented as bit strings called chromosomes (or individuals) and then evolved. In the scheduling problem solved by Pezzella et al. [14], there were 8 operations. Therefore, a feasible schedule was represented using a chromosome diagram containing 8 genes (see Fig. 4.2). For example, the first operation in this figure was (1, 1, 3), which meant that for job #1, its first operation was performed on machine #3. However, different job scheduling problems require different encodings, and some may be complex.

A Gantt chart for visualizing the schedule is shown in Fig. 4.3. The makespan of the schedule is derived using a simplified production simulator as 13.

Assume that each population contains Q schedules/chromosomes. The schedules of the first population can be randomly generated (see Fig. 4.4). A simplified production simulator is also used to derive the makespan of each schedule. The fitness of the schedule is then set to the reciprocal of the makespan. According to the fitness proportion **selection** principle, to create a new intermediate population of Q "parents," Q independent extractions need to be made from the old population. The probability of selecting a schedule is linearly proportional to its fitness. For example, in Fig. 4.4, the frequency that the first schedule is chosen can be calculated as

Min $Z = C_{max}$

s.t.

$$C_{max} \geq C_j ; j = 1 \sim 3$$

$$C_1 = c_{13} \qquad \qquad ' c_{jk} , \, k: \text{ the no. of processing step}$$

$$C_2 = c_{23}$$
$$C_3 = c_{32}$$

$$c_{11} = s_{11} + X_{111} \cdot 7 + X_{112} \cdot 6 + X_{113} \cdot 4 + X_{114} \cdot 5 \qquad ' s_{jk} , \; X_{jkm} , \, m: \text{ the machine no.}$$

$$X_{111} + X_{112} + X_{113} + X_{114} = 1$$

$$c_{12} = s_{12} + X_{121} \cdot 4 + X_{122} \cdot 8 + X_{123} \cdot 5 + X_{124} \cdot 6$$

$$X_{121} + X_{122} + X_{123} + X_{124} = 1$$

$$c_{13} = s_{13} + X_{131} \cdot 9 + X_{132} \cdot 5 + X_{133} \cdot 4 + X_{134} \cdot 7$$

$$X_{131} + X_{132} + X_{133} + X_{134} = 1$$

$$s_{12} \geq c_{11}$$
$$s_{13} \geq c_{12}$$

$$\ldots$$

$$s_{21} \geq (X_{111} \cdot X_{211} + X_{112} \cdot X_{212} + X_{113} \cdot X_{213} + X_{114} \cdot X_{214}) Y_{1121} \cdot c_{11}$$

$$s_{11} \geq (X_{111} \cdot X_{211} + X_{112} \cdot X_{212} + X_{113} \cdot X_{213} + X_{114} \cdot X_{214}) Y_{2111} \cdot c_{21}$$

$$Y_{1121} + Y_{2111} = 1$$

$$s_{31} \geq (X_{111} \cdot X_{311} + X_{112} \cdot X_{312} + X_{113} \cdot X_{313} + X_{114} \cdot X_{314}) Y_{1131} \cdot c_{11}$$

$$s_{11} \geq (X_{111} \cdot X_{311} + X_{112} \cdot X_{312} + X_{113} \cdot X_{313} + X_{114} \cdot X_{314}) Y_{3111} \cdot c_{31}$$

$$Y_{1131} + Y_{3111} = 1$$

$$\ldots$$

$$X_{jkm} , \; Y_{j_1 k_1 j_2 k_2} \; \in \{0, 1\} \qquad ' Y_{j_1 k_1 j_2 k_2} = 1 \text{ if } (j_1, k_1) \rightarrow (j_2, k_2)$$

Other variables $\in R^+$

Fig. 4.1 MINLP problem for the flexible job shop scheduling problem

| (1, 1, 3) | (1, 2, 1) | (2, 1, 3) | (2, 2, 4) | (3, 1, 3) | (1, 3, 2) | (3, 2, 1) | (2, 3, 4) |

Fig. 4.2 Chromosome for representing a feasible schedule

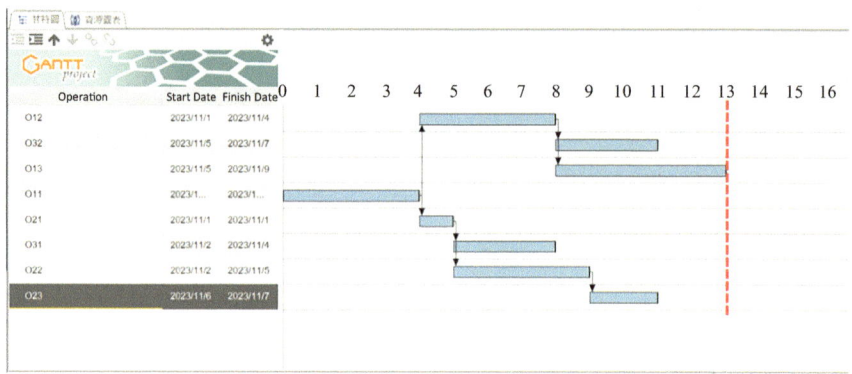

Fig. 4.3 Gantt chart for visualizing the schedule

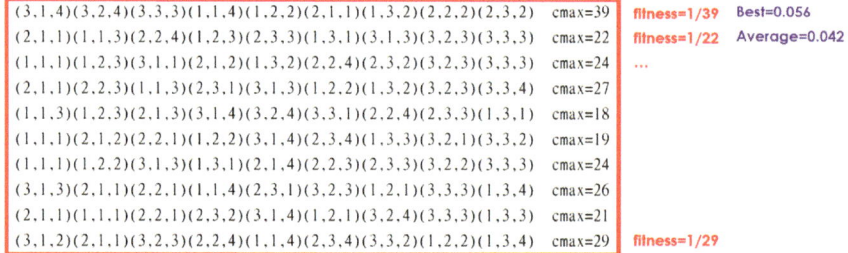

Fig. 4.4 Schedules of a population

$$10 \cdot \frac{\frac{1}{39}}{\frac{1}{39} + \frac{1}{22} + \cdots + \frac{1}{29}} = 0.61$$

In this way, above-average schedules will have more copies in new populations, while below-average schedules will be at risk of extinction.

Couples are formed with all parent schedules. Every couple experiences crossover. With a certain probability (called the crossover rate p_{cross}), the two parent schedules are cut at the same random position, and the remaining parts are swapped between the two parent schedules (see Fig. 4.5). Any redundant or ineligible operation is randomly replaced by an unassigned and eligible operation on a random machine (see Fig. 4.6):

- Redundancy: The same operation has been performed on any machine.
- Ineligibility: The previous operation has not been performed yet.

In this way, two novel offspring schedules are yielded, each containing operations from both parent schedules.

Subsequently, mutation is to simulate the effect of transcription errors that can happen with a very low probability (p_{mut}) when a schedule is duplicated. This is

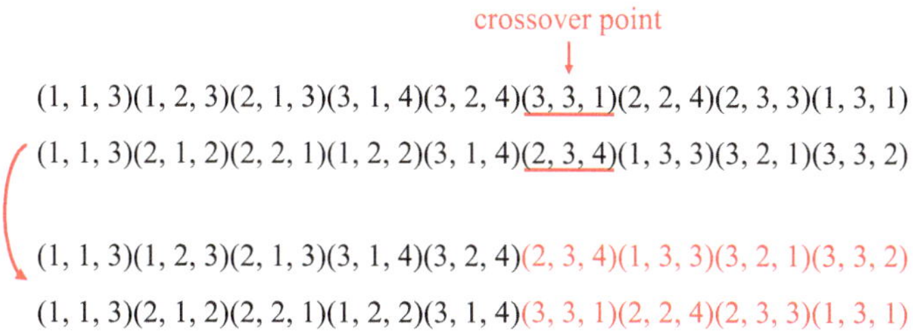

crossover point
↓

(1, 1, 3)(1, 2, 3)(2, 1, 3)(3, 1, 4)(3, 2, 4)(3, 3, 1)(2, 2, 4)(2, 3, 3)(1, 3, 1)

(1, 1, 3)(2, 1, 2)(2, 2, 1)(1, 2, 2)(3, 1, 4)(2, 3, 4)(1, 3, 3)(3, 2, 1)(3, 3, 2)

(1, 1, 3)(1, 2, 3)(2, 1, 3)(3, 1, 4)(3, 2, 4)(2, 3, 4)(1, 3, 3)(3, 2, 1)(3, 3, 2)

(1, 1, 3)(2, 1, 2)(2, 2, 1)(1, 2, 2)(3, 1, 4)(3, 3, 1)(2, 2, 4)(2, 3, 3)(1, 3, 1)

Fig. 4.5 Crossover operation

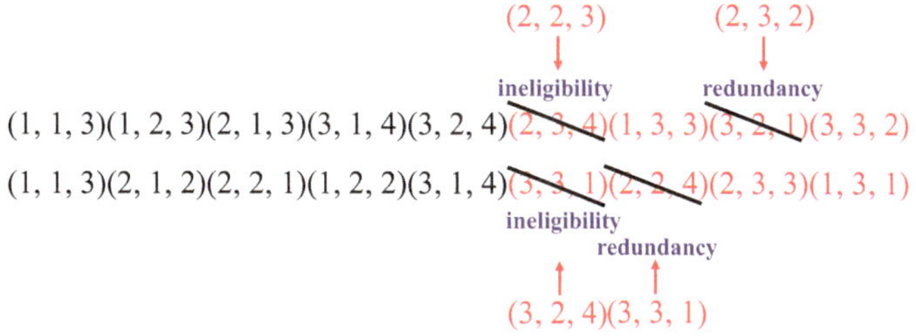

(2, 2, 3)   (2, 3, 2)
↓   ↓
ineligibility   redundancy

(1, 1, 3)(1, 2, 3)(2, 1, 3)(3, 1, 4)(3, 2, 4)(2, 3, 4)(1, 3, 3)(3, 2, 1)(3, 3, 2)

(1, 1, 3)(2, 1, 2)(2, 2, 1)(1, 2, 2)(3, 1, 4)(3, 3, 1)(2, 2, 4)(2, 3, 3)(1, 3, 1)

ineligibility
redundancy
↑   ↑
(3, 2, 4)(3, 3, 1)

Fig. 4.6 Solving redundancy and ineligibility

accomplished by changing the machine on which an operation is performed (see Fig. 4.7).

Common termination conditions include

- A predetermined number of generations or time has elapsed.
- A satisfactory solution has been found.
- No improvement in solution quality has taken place for a predetermined number of generations (see Fig. 4.8).

A comparison of the understandability of various technologies is as follows:

- MP models used to solve job sequencing and scheduling problems are easy to interpret, such as checking the feasibility of a schedule or the correctness of the derived scheduling performance.
- Traditional optimization techniques (such as branch and bound) or software for solving MP models are black boxes and difficult to understand.

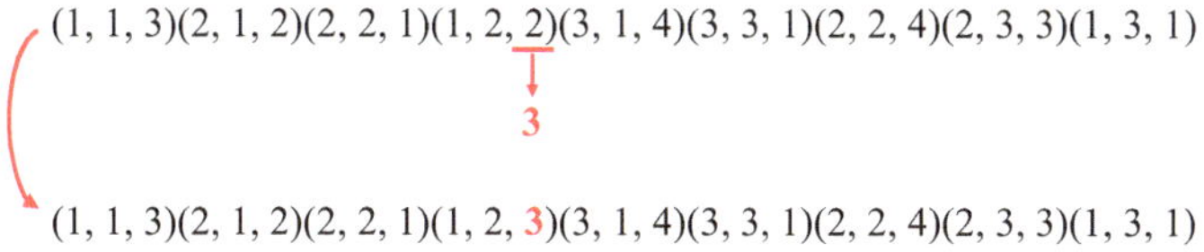

(1, 1, 3)(2, 1, 2)(2, 2, 1)(1, 2, 2)(3, 1, 4)(3, 3, 1)(2, 2, 4)(2, 3, 3)(1, 3, 1)

3

(1, 1, 3)(2, 1, 2)(2, 2, 1)(1, 2, 3)(3, 1, 4)(3, 3, 1)(2, 2, 4)(2, 3, 3)(1, 3, 1)

Fig. 4.7 Mutation operation

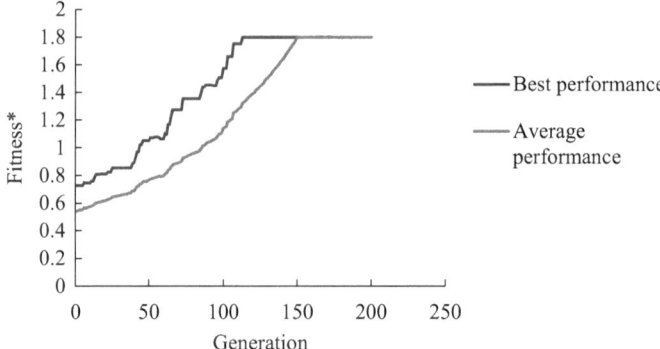

Fig. 4.8 Dynamic line chart for tracing the convergence of the optimal solution

- The GA that helps solve a MP problem is also a black box and difficult to understand.

4.3 XAI Techniques and Tools for Explaining GA Applications in Job Sequencing and Scheduling

The implementation process of the GA is shown in Fig. 4.9 [17]. XAI techniques and tools can be applied to several steps of the process.

In addition, XAI techniques and tools for explaining GA can be easily extended to account for other evolutionary AI applications such as ant colony optimization (ACO), particle swarm optimization (PSO), and artificial bee colony (ABC) applications in job sequencing and scheduling.

Existing XAI techniques for explaining GA applications in job sequencing and scheduling are subject to the following problems:

- In a GA application for job sequencing and scheduling, the inputs are usually feasible solutions rather than job attributes, and the output is the optimal schedule [15]. However, there is no direct relationship between feasible and optimal solutions, because if the initial feasible solutions are different, the optimal solution may be still the same. Therefore, prevalent methods for fitting input–output relationships, such as decision trees [18], fuzzy inference rules (or systems), random forests (RFs) [19], classification and regression trees (CART) [19], Shapley additive explanations (SHAP) [1], and local interpretable model-agnostic explanation (LIME) [20] are not suitable for explaining GA applications in job sequencing and scheduling.
- In addition, most of the popular input–output relationship fitting methods are difficult to deal with a huge number of inputs [21], just like GA applications [22].

Fig. 4.9 Implementation process of a GA with XAI applications

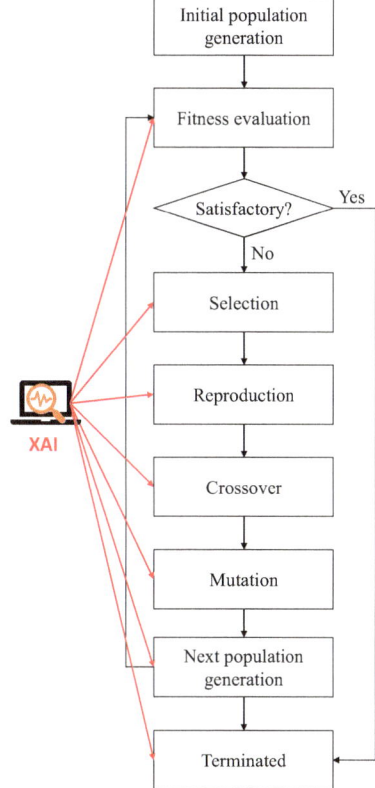

• Furthermore, in previous studies, chromosome maps are usually used to describe the evolution of feasible solutions. How these changes lead to the optimal solution has not yet been visualized.

4.3.1   Textual Descriptions-Based XAI Techniques

Text descriptions are the most common technique for explaining GA applications and may reduce the user's need for further explanations of the GA application. An example is given in Fig. 4.10.

Pseudocode (Fig. 4.11) is also similar to textual description [23], but may be better understood by users with basic programming training, which is easier for engineers in a factory.

Another textual technique for explaining GA is a **flowchart** (see Fig. 4.9), which is also a visual tool but uses abstract text in a variety of boxes to convey the concepts. For factory personnel, flowcharts are not necessarily more difficult to understand than pseudocodes.

First, the encoding of a chromosome is illustrated in Fig. 4, where 0 represents selecting the left α cut of a fuzzy pairwise comparison result; 1 represents selecting the right α cut. There are two fitness functions. One is to maximize fuzzy maximal eigenvalue (or fuzzy vector), and the other is to minimize fuzzy maximal eigenvalue (or fuzzy vector):

$$\text{Max } fitness = -\lambda^L(\alpha) \text{ (or } -\mathbf{x}^L(\alpha)) \tag{12}$$

$$\text{Max } fitness = \lambda^R(\alpha) \text{ (or } \mathbf{x}^R(\alpha)) \tag{13}$$

The optimization results are used to establish the α cut of fuzzy eigenvalue (or fuzzy vector). To this end, two groups of chromosomes are established. Because the two fitness functions are opposite, chromosomes that perform particularly poorly in one group can be moved to the other.

The roulette wheel method is applied to choose parent chromosomes to be paired based on their fitness values. A crossover point is chosen at random. Offspring chromosomes are generated by exchanging the genes of parents among themselves before or after the crossover point.

Fig. 4.10   Textual description for explaining a GA application

```
nchromo = number of chromosomes in a population;
pc = crossover probability;
pm = mutation probability;
T = maximum number of populations;
popu(t) = the t-th population;
t = the current population no.;

t = 1;
Generate the initial population popu(1);
while t < T do
    rdcr, rdm1, rdm2 = random numbers between 0 and 1;
    chromo1, chromo2 = chromosome in popu(t);
    Remove chromo1, chromo2 from popu(t);
    if rdcr < pc
     chromo3, chromo4 = crossover chromo1 and chromo2;
    end if
    if rdm1 < pm
     mutate chromo3;
    end if
    if rdm2 < pm
     mutate chromo4;
    end if
    Select two chromosomes from    chromo1 ~  chromo4 with the highest fitness and add to
    popu(t+1);
    t++;
end while
```

Fig. 4.11   Pseudocode for GA

These three text description techniques are commonly used to generate **global explanations** about the reasoning mechanism of GA.

4.3.2   Generic Visualization Techniques

Generic visualization techniques have been applied to explain GA applications for job scheduling. For example, a **system architecture diagram** (Fig. 4.12) can illustrate the scheduling system that incorporates a GA application [8].

A **flowchart** can also be drawn to detail the steps involved in applying the GA (see Fig. 4.9).

Chromosome diagrams (or diagrams) are the most popular visualization tool used to illustrate the structure of chromosomes (see Fig. 4.2). Chromosome diagrams (or diagrams) are also used to explain the steps of a GA application, such as crossover and mutation (see Figs. 4.5, 4.6 and 4.7).

One possible way to visualize chromosomes is to use separate colors to express the operations of different jobs within them. If there are n jobs, n different colors should be chosen. These colors should be as different from each other as possible to make them easier for users to distinguish. To this end, the following mixed integer-nonlinear programming problem (MINP) can be solved:

$$\text{Max } Z = \min_{i \neq j} d_{ij} \tag{4.1}$$

s.t.

$$d_{ij} = \sqrt{(R_i - R_j)^2 + (G_i - G_j)^2 + (B_i - B_j)^2}; i, j = 1 \sim n \tag{4.2}$$

Fig. 4.12 System architecture diagram highlighting a GA application

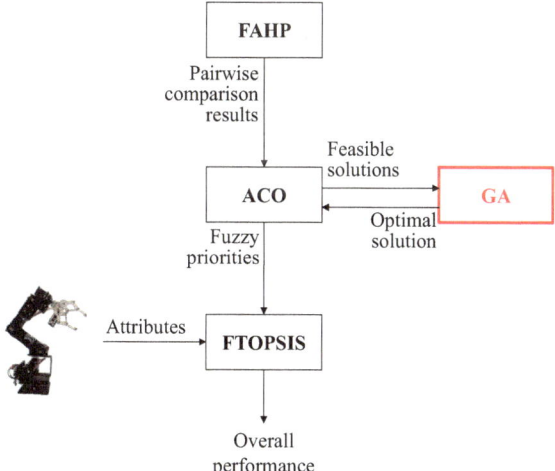

$$R_j \leq 255 j = 1 \sim n \tag{4.3}$$

$$G_j \leq 255 j = 1 \sim n \tag{4.4}$$

$$B_j \leq 255 j = 1 \sim n \tag{4.5}$$

$$R_j, G_j, B_j \in \mathbf{Z}^+; j = 1 \sim n \tag{4.6}$$

Take the chromosome in Fig. 4.2 as an example. $n = 3$. Lingo is applied to solve the MINP problem (see Fig. 4.13). As expected, the optimal solution is

$$\text{Job \#1: } (R_1, G_1, B_1) = (0, 0, 255)$$

$$\text{Job \#2: } (R_2, G_2, B_2) = (255, 0, 0)$$

$$\text{Job \#3: } (R_1, G_1, B_1) = (0, 255, 0)$$

as shown in Fig. 4.14. If there are a lot of jobs, choosing different colors for them becomes a challenging task.

Subsequently, in order to distinguish the operations assigned to different machines, each bar can be divided into two parts. The upper half maintains the color of the job, while the lower half represents the machine number in varying shades (Fig. 4.15).

Another treatment is to use the depth of the bar color to show the processing time required for an operation:

```
max=Z;
Z<=d12;
Z<=d13;
Z<=d23;
d12^2=(R1-R2)^2+(G1-G2)^2+(B1-B2)^2;
d13^2=(R1-R3)^2+(G1-G3)^2+(B1-B3)^2;
d23^2=(R2-R3)^2+(G2-G3)^2+(B2-B3)^2;
R1<=255;
R2<=255;
R3<=255;
G1<=255;
G2<=255;
G3<=255;
B1<=255;
B2<=255;
B3<=255;
@GIN(R1);@GIN(R2);@GIN(R3);@GIN(B1);@GIN(B2);@GIN(B3);@GIN(G1);@GIN(G2);@GIN
(G3);
```

Fig. 4.13 Lingo code for solving the MINP problem

Fig. 4.14   Color encoding of chromosomes

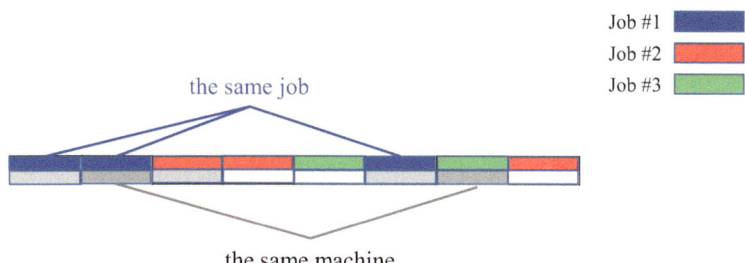

Fig. 4.15   Color encoding including machine numbers

$$R_{jk} = \max\left(R_j + 255 - \frac{p_{jk}}{\max\limits_{l,m} p_{lm}} \cdot 255, 255\right) j = 1 \sim n \qquad (4.7)$$

$$G_{jk} = \max\left(G_j + 255 - \frac{p_{jk}}{\max\limits_{l,m} p_{lm}} \cdot 255, 255\right) j = 1 \sim n \qquad (4.8)$$

$$B_{jk} = \max\left(B_j + 255 - \frac{p_{jk}}{\max\limits_{l,m} p_{lm}} \cdot 255, 255\right) j = 1 \sim n \qquad (4.9)$$

The result is illustrated in Fig. 4.16.

During the evolution/optimization process, a **dynamic line chart** can be used to trace the convergence of the optimal solution [17] (see Fig. 4.8). The feasible solutions (schedules) of the last population may be very similar (see Fig. 4.17).

The best tool for visualizing the scheduling results is a **Gantt chart** (see Fig. 4.3).

Fig. 4.16   Color encoding considering processing times

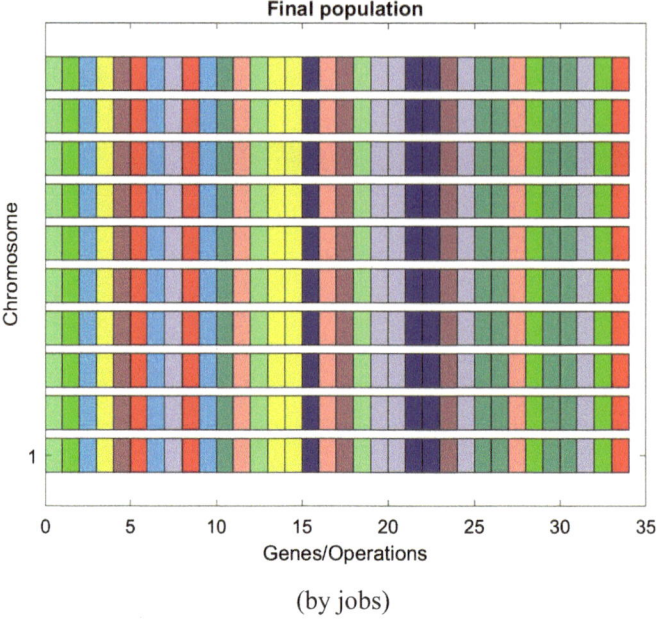

(by jobs)

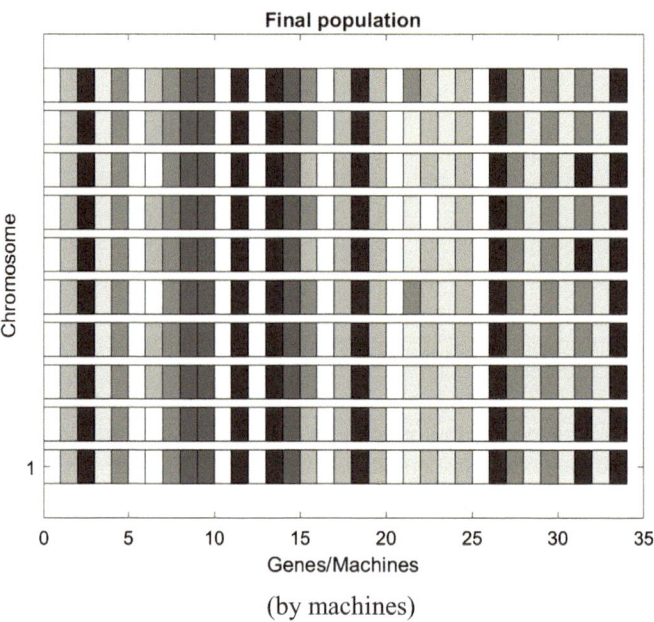

(by machines)

Fig. 4.17 Feasible solutions (schedules) of the last population may

Fig. 4.18 Bar chart for comparing scheduling performances of various scheduling methods

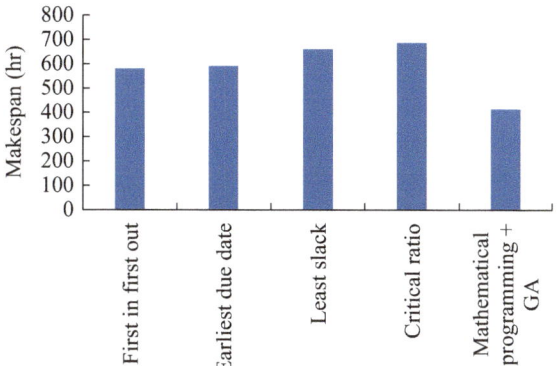

After optimization, a **bar chart** can be used to compare the scheduling performances of various scheduling methods [19] (see Fig. 4.18).

4.3.3   Saliency Maps

Saliency maps are images that highlight the areas where people's eyes focus first or which areas are most relevant to a machine learning model [24].

To build a saliency map for a frame, first, the distance of each pixel to the rest of pixels in the frame is calculated:

$$\text{SALS}(I_k) = \sum_{i=1}^{n} |I_k - I_i| \tag{4.10}$$

$I_i \in [0, 255]$ is the value of pixel; n is the number of pixels in the frame. To highlight pixels more distinct from the others,

$$\text{SALS}(I_k) = \sum_{l} F_l |I_k - I_l| \tag{4.11}$$

F_l is the frequency of I_l. $l \in [0, 255]$.

The concept of saliency maps can be applied to associate points in the dynamic line chart with different colors to distinguish their performances in several ways. First, data points can be associated with different colors according to their performances:

$$\left[R_{p_t}, G_{p_t}, B_{p_t}\right] = \left[255 \cdot \left(1 - \frac{p_t - \min_s p_s}{\max_s p_s - \min_s p_s}\right), 255 \cdot \frac{p_t - \min_s p_s}{\max_s p_s - \min_s p_s}, 0\right] \tag{4.12}$$

where p_t is the performance/fitness of population t. In Fig. 4.19, green points are associated with better scheduling performances than red points. The required MATLAB code is provided in Fig. 4.20. Thus, when a data point turns green, it indicates that the performance is above the average. Whether the evolution process needs to be terminated depends on how green the current data point is for the scheduler.

Contrastive gradient-based saliency maps [25] are a special type of saliency maps. The concept of contrastive gradient-based saliency maps can also be applied to the dynamic line chart. This method calculates the saliency map for a specific input x_t. The map is usually calculated as the gradient of the output (o_t) with respect to the input:

$$\frac{\partial o_t}{\partial x_t} = \lim_{\Delta t \to 0} \frac{o_{t+\Delta t} - o_t}{\Delta t} \tag{4.13}$$

By definition, the rate of change in the output with respect to the input is the gradient of the output with respect to the input and it represents how much will the output change with a small change in the input. In other words, these gradients represent how stable the machine learning (ML) or deep learning (DL) algorithm is [26]. It is important to note that this gradient is calculated during the training process. Therefore, after applying the concept of contrastive gradient-based saliency maps to the dynamic line chart, the color associated with a data point becomes

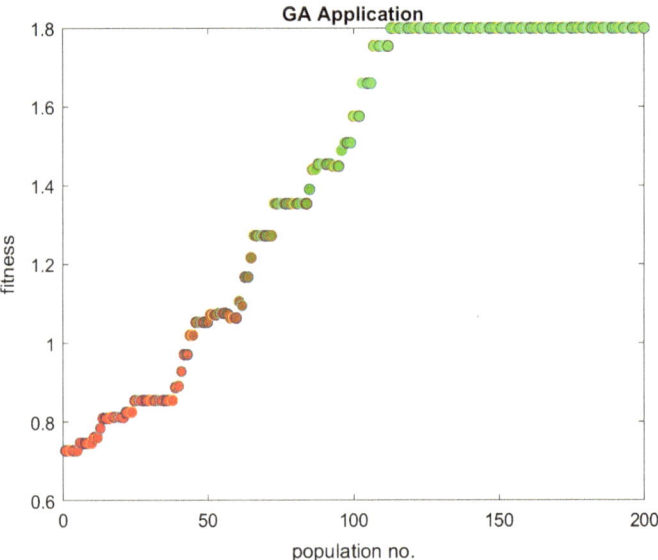

Fig. 4.19 Saliency map application to the dynamic line chart

```
bestperformance=GA3(:,2);
averageperformance=GA3(:,3);
iteration=GA3(:,1);

plot(iteration(1,1),bestperformance(1,1),'o','MarkerFaceColor',[1-(bestperformance(1,1)-
min(bestperformance(:,1)))/(max(bestperformance(:,1))-min(bestperformance(:,1)))
(bestperformance(1,1)-min(bestperformance(:,1)))/(max(bestperformance(:,1))-
min(bestperformance(:,1))) 0]);
hold;

for i=2:200
    plot(iteration(i,1),bestperformance(i,1),'o','MarkerFaceColor',[1-(bestperformance(i,1)-
min(bestperformance(:,1)))/(max(bestperformance(:,1))-min(bestperformance(:,1)))
(bestperformance(i,1)-min(bestperformance(:,1)))/(max(bestperformance(:,1))-
min(bestperformance(:,1))) 0]);
end

xlabel('population no.');
ylabel('fitness');
title('GA Application');
```

Fig. 4.20 Required MATLAB code

$$R_{p_t} = 255 \cdot \frac{\frac{p_{t+\Delta t}-p_t}{\Delta t} - \min_s\left(\frac{p_{s+\Delta t}-p_s}{\Delta t}\right)}{\max_s\left(\frac{p_{s+\Delta t}-p_s}{\Delta t}\right) - \min_s\left(\frac{p_{s+\Delta t}-p_s}{\Delta t}\right)} \tag{4.14}$$

$$G_{p_t} = 255 \cdot \left(1 - \frac{\frac{p_{t+\Delta t}-p_t}{\Delta t} - \min_s\left(\frac{p_{s+\Delta t}-p_s}{\Delta t}\right)}{\max_s\left(\frac{p_{s+\Delta t}-p_s}{\Delta t}\right) - \min_s\left(\frac{p_{s+\Delta t}-p_s}{\Delta t}\right)}\right) \tag{4.15}$$

$$B_{p_t} = 0 \tag{4.16}$$

as illustrated in Fig. 4.21. The required MATLAB code is provided in Fig. 4.22. However, in GA applications, it is possible for successive populations to have the same optimal performance. Therefore, Δt is set to a slightly higher value, such as 50, in Fig. 4.23.

The distinguishing effect can be further strengthened as follows:

$$R_{p_t} = 255 \cdot \left(\frac{\frac{p_{t+\Delta t}-p_t}{\Delta t} - \min_s\left(\frac{p_{s+\Delta t}-p_s}{\Delta t}\right)}{\max_s\left(\frac{p_{s+\Delta t}-p_s}{\Delta t}\right) - \min_s\left(\frac{p_{s+\Delta t}-p_s}{\Delta t}\right)}\right)^{\varphi} \tag{4.17}$$

$$G_{p_t} = 255 \cdot \left(1 - \left(\frac{\frac{p_{t+\Delta t}-p_t}{\Delta t} - \min_s\left(\frac{p_{s+\Delta t}-p_s}{\Delta t}\right)}{\max_s\left(\frac{p_{s+\Delta t}-p_s}{\Delta t}\right) - \min_s\left(\frac{p_{s+\Delta t}-p_s}{\Delta t}\right)}\right)^{\varphi}\right) \tag{4.18}$$

$$B_{p_t} = 0 \tag{4.19}$$

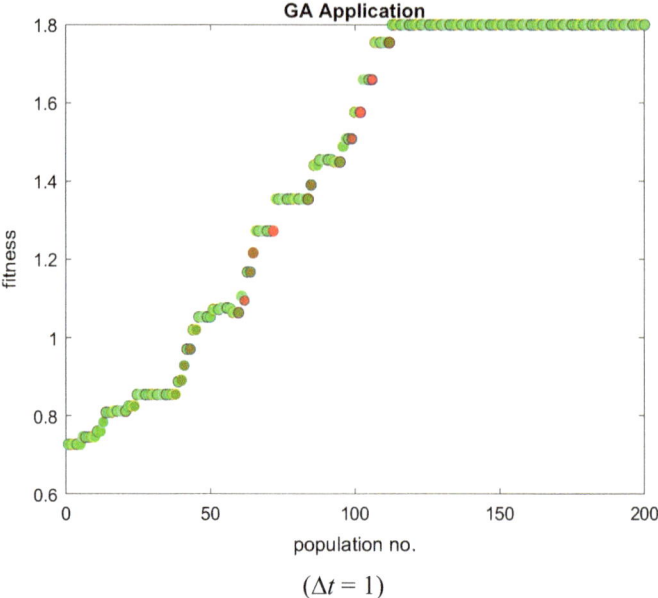

$$(\Delta t = 1)$$

Fig. 4.21 Contrastive gradient-based saliency map application to the dynamic line chart

```
bestperformance=GA3(:,2);
averageperformance=GA3(:,3);
iteration=GA3(:,1);
dt=50;
grad=zeros(200,1);
for i=1:200-dt
    grad(i,1)=(bestperformance(i+dt,1)-bestperformance(i,1))/dt;
end

grad=grad.^(1/4);

plot(iteration(1,1),bestperformance(1,1),'o','MarkerFaceColor',[(grad(1,1)-
min(grad(:,1)))/(max(grad(:,1))-min(grad(:,1))) 1-(grad(1,1)-min(grad(:,1)))/(max(grad(:,1))-
min(grad(:,1))) 0]);
hold;

for i=2:200
    plot(iteration(i,1),bestperformance(i,1),'o','MarkerFaceColor',[(grad(i,1)-
min(grad(:,1)))/(max(grad(:,1))-min(grad(:,1))) 1-(grad(i,1)-min(grad(:,1)))/(max(grad(:,1))-
min(grad(:,1))) 0]);
end

xlabel('population no.');
ylabel('fitness');
title('GA Application');
```

Fig. 4.22 MATLAB code

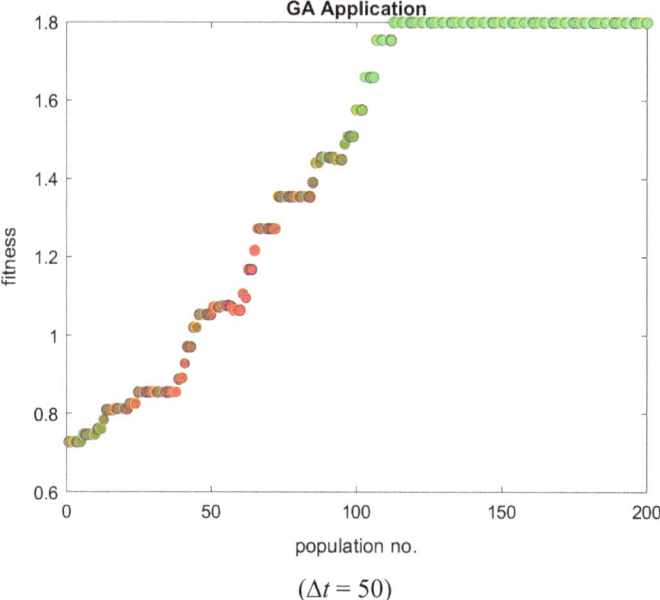

$$(\Delta t = 50)$$

Fig. 4.23 Contrastive gradient-based saliency map application by setting Δt to a slightly higher value

as illustrated in Fig. 4.24, where $\varphi = 0.25$.

The concept of contrastive gradient-based saliency maps can also be applied to highlight the most important operations (using red bars) of the Gantt chart in Fig. 4.3. In contrast, greener bars indicate less important operations to minimize the makespan. For example, the gradient of the makespan/output with respect to the start time/input of each operation in the optimal solution can be calculated:

$$I_{O_l} = \frac{\partial C_{\max}}{\partial s_{O_l}} = \lim_{\Delta s_{O_l} \to 0} \frac{C_{\max}\left(s_{O_l} + \Delta s_{O_l}\right) - C_{\max}\left(s_{O_l}\right)}{\Delta s_{O_l}} \tag{4.20}$$

Then,

$$R_{p_t} = 255 \cdot \frac{\frac{C_{\max}\left(s_{O_l} + \Delta s_{O_l}\right) - C_{\max}\left(s_{O_l}\right)}{\Delta s_{O_l}} - \min_q\left(\frac{C_{\max}\left(s_{O_q} + \Delta s_{O_q}\right) - C_{\max}\left(s_{O_q}\right)}{\Delta s_{O_q}}\right)}{\max_q\left(\frac{C_{\max}\left(s_{O_q} + \Delta s_{O_q}\right) - C_{\max}\left(s_{O_q}\right)}{\Delta s_{O_q}}\right) - \min_q\left(\frac{C_{\max}\left(s_{O_q} + \Delta s_{O_q}\right) - C_{\max}\left(s_{O_q}\right)}{\Delta s_{O_q}}\right)} \tag{4.21}$$

$$G_{p_t} = 255 \cdot \left(1 - \frac{\frac{C_{\max}\left(s_{O_l} + \Delta s_{O_l}\right) - C_{\max}\left(s_{O_l}\right)}{\Delta s_{O_l}} - \min_q\left(\frac{C_{\max}\left(s_{O_q} + \Delta s_{O_q}\right) - C_{\max}\left(s_{O_q}\right)}{\Delta s_{O_q}}\right)}{\max_q\left(\frac{C_{\max}\left(s_{O_q} + \Delta s_{O_q}\right) - C_{\max}\left(s_{O_q}\right)}{\Delta s_{O_q}}\right) - \min_q\left(\frac{C_{\max}\left(s_{O_q} + \Delta s_{O_q}\right) - C_{\max}\left(s_{O_q}\right)}{\Delta s_{O_q}}\right)}\right) \tag{4.22}$$

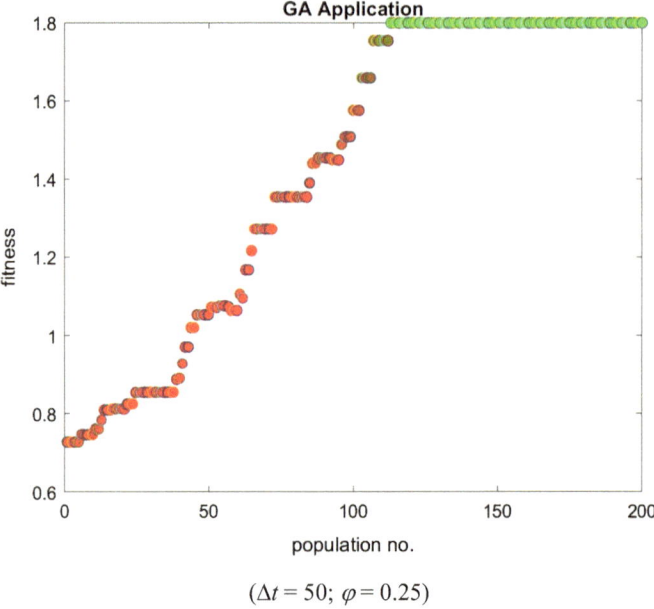

$$(\Delta t = 50; \ \varphi = 0.25)$$

Fig. 4.24 Strengthening the distinguishing effect

$$B_{p_t} = 0 \tag{4.23}$$

The result is shown in Fig. 4.25, where $\Delta s_{o_l} = 1; l = 1 \sim 8$. Red operations are more important than green operations. Figure 4.25 is in line with the concept of emphasizing critical chain operations since all red bars are on the critical chain [27].

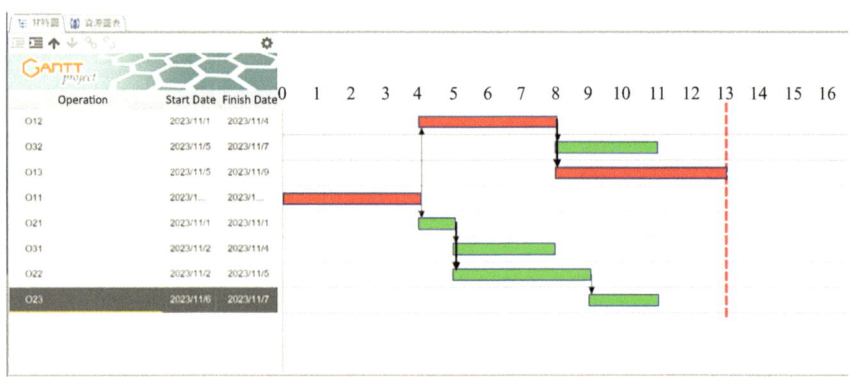

Fig. 4.25 Contrastive gradient-based saliency map application to the Gantt chart

For other operations, their importance levels are evaluated with the changes that should be made to their start times to worsen the scheduling performance. For this purpose, Eq. (4.20) is changed to

$$I_{o_l} = \frac{\Delta C_{\max}}{\Delta s_{o_l}} = \frac{C_{\max}(s_{o_l} + \Delta s_{o_l}) - C_{\max}(s_{o_l})}{\Delta s_{o_l}} = \frac{C_{\max}(s_{o_l}) + 1 - C_{\max}(s_{o_l})}{\Delta s_{o_l}} = \frac{1}{\Delta s_{o_l}} \tag{4.24}$$

The importance levels of operations in Fig. 4.3 are evaluated and summarized in Table 4.1.

The color associated with an operation is determined as follows:

$$R_{p_t} = 255 \cdot \frac{I_{o_l} - \min_q(I_{o_q})}{\max_q(I_{o_q}) - \min_q(I_{o_q})} \tag{4.25}$$

$$G_{p_t} = 255 \cdot \left(1 - \frac{I_{o_l} - \min_q(I_{o_q})}{\max_q(I_{o_q}) - \min_q(I_{o_q})}\right) \tag{4.26}$$

$$B_{p_t} = 0 \tag{4.27}$$

To further distinguish the importance levels,

$$R_{p_t} = 255 \cdot \left(\frac{I_{o_l} - \min_q(I_{o_q})}{\max_q(I_{o_q}) - \min_q(I_{o_q})}\right)^{\varphi} \tag{4.28}$$

$$G_{p_t} = 255 \cdot \left(1 - \left(\frac{I_{o_l} - \min_q(I_{o_q})}{\max_q(I_{o_q}) - \min_q(I_{o_q})}\right)^{\varphi}\right). \tag{4.29}$$

Table 4.1 Operation importance evaluation results

l	j (Job)	o (Operation)	m (Machine)	I_{o_l}
1	1	2	1	1
2	3	2	1	1/3
3	1	3	2	1
4	1	1	3	1
5	2	1	3	1/3
6	3	1	3	1/6
7	2	2	4	1/3
8	2	3	4	1/3

Table 4.2 Colors of operations

l	j (Job)	o (Operation)	m (Machine)	R	G	B
1	1	2	1	255	0	0
2	3	2	1	0	255	0
3	1	3	2	137	118	0
4	1	1	3	255	0	0
5	2	1	3	118	137	0
6	3	1	3	0	255	0
7	2	2	4	59	196	0
8	2	3	4	0	255	0

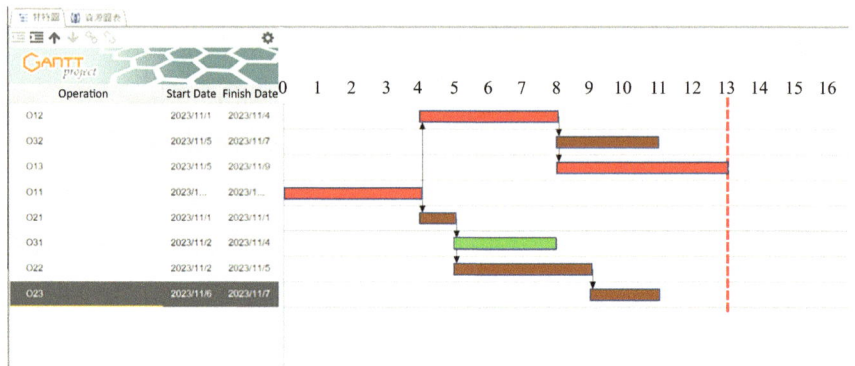

Fig. 4.26 Evaluation results by considering the direct impacts on other jobs

$$B_{p_t} = 0 \qquad\qquad (4.30)$$

For the example in Table 4.1, the results ($\varphi = 0.25$) are summarized in Table 4.2 and illustrated in Fig. 4.26.

4.3.4   Decision Tree-Based Interpretation

Decision trees are a highly interpretable XAI technique. A decision rule has the following format [28, 29]:

$$\text{"If } x_{j1} \wedge_{q1} c_{q1} \text{ and } \ldots \text{ and } x_{jP} \wedge_{qP} c_{qP} \text{ then } \hat{o}_j = d_q, \text{"} \qquad (4.31)$$

where c_{qp} is the threshold for x_{jp} in the q-th decision rule; $q = 1 \sim Q$ (the number of rules). $\wedge_{qp} \in \{=, <, \geq, \text{within}\}$ is the used logical operator. $\hat{o}_j$ is the approximated output. d_q is the outcome of decision rule q.

Each decision tree is trained to minimize the RMSE of approximating $\hat{o}_j$ with o_j:

$$\text{RMSE} = \sqrt{\frac{\sum_{j=1}^{n} (\hat{o}_j - o_j)^2}{n}} \qquad (4.32)$$

subject to constraints such as the minimum number of jobs of a branch, the maximum number of nodes, and the tallest tree height/depth [30]. The process of building each decision tree goes through three phases: tree growth, stopping, and pruning.

Decision trees have been constructed to explain AI applications in many domains such as estimation, prediction, and pattern recognition. Wang and Chen [31] explained the operations of GA applications in job sequencing and scheduling with decision trees, as detailed in the following.

First, the selection mechanism can be explained by the decision tree in Fig. 4.27, in which $C_{\max,q}$ is the makespan (i.e., the reciprocal of fitness) associated with chromosome q; $q = 1 \sim Q$. $\{\mathbf{x}_s\}$ is the set of chromosomes of the current population. From $\{\mathbf{x}_s\}$, paired chromosomes $\{\mathbf{x}_p\}$ are to be selected. l is the index of a gene/operation. For example, $\mathbf{x}_s = \{x_{sl}\}$; $x_{sl} = (j_{sl}, o_{sl}, m_{sl})$. Other selection mechanisms can be explained similarly.

The crossover operation in Fig. 4.5 can then be interpreted in terms of the decision tree in Fig. 4.28, where p and p' indicate the two parental chromosomes crossed over, which results in the two offspring chromosomes o and o'. Other crossover mechanisms can be explained similarly.

Subsequently, the mutation operation in Fig. 4.7 is explained using another decision tree, as shown in Fig. 4.29, where the machine of the operation in position l is changed. Other mutational mechanisms can be explained similarly.

These decision trees can be combined to explain the complete evolutionary process of GA, as shown in Fig. 4.30.

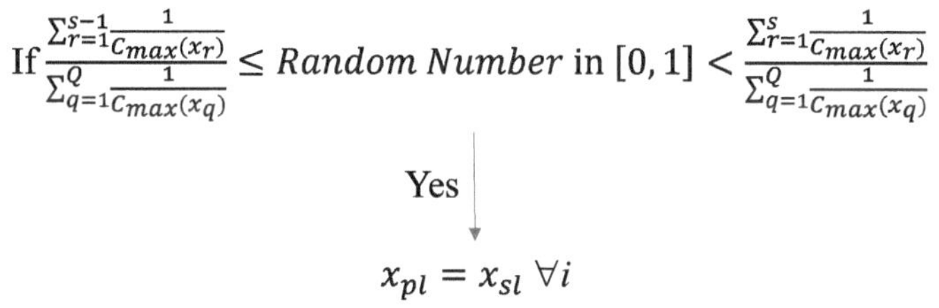

$$\text{If } \frac{\sum_{r=1}^{s-1} \frac{1}{C_{max}(x_r)}}{\sum_{q=1}^{Q} \frac{1}{C_{max}(x_q)}} \leq Random\ Number \text{ in } [0,1] < \frac{\sum_{r=1}^{s} \frac{1}{C_{max}(x_r)}}{\sum_{q=1}^{Q} \frac{1}{C_{max}(x_q)}}$$

Yes

$$x_{pl} = x_{sl}\ \forall i$$

Fig. 4.27 Decision tree for explaining the selection mechanism

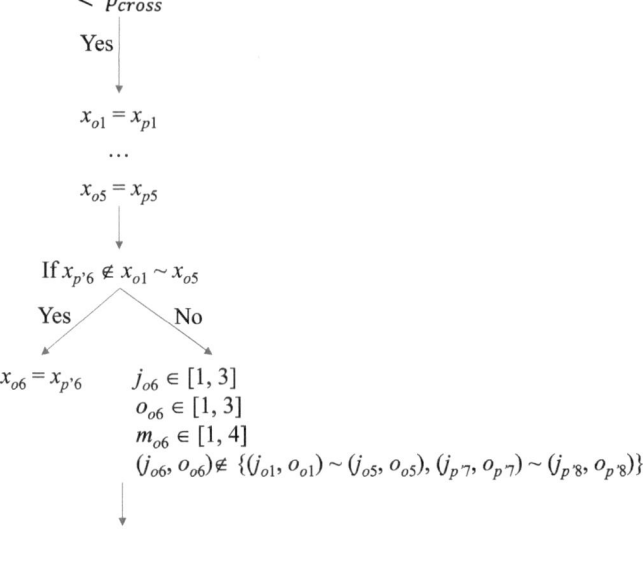

Fig. 4.28 Decision tree for explaining the crossover mechanism

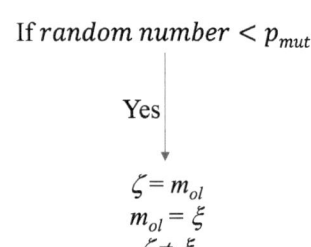

Fig. 4.29 Decision tree for
explaining the mutation
mechanism

4.3.5 Dynamic Transition and Contribution Diagram

Chromosome diagrams should be read and cannot be comprehended at once. Chromosome diagrams of different schedules also look similar. In addition, Gantt chart will be difficult to draw and view if there are many machines, jobs, and steps. To address these issues, Wang and Chen [31] designed a new XAI tool, radar charts, to provide a convenient way to visualize schedule differences. Take the following schedule as an example,

$$(1, 1, 3)(1, 2, 1)(2, 1, 3)(2, 2, 4)(3, 1, 3)(1, 3, 2)(3, 2, 1)(2, 3, 4).$$

It is visualized as a radar chart in Fig. 4.31.

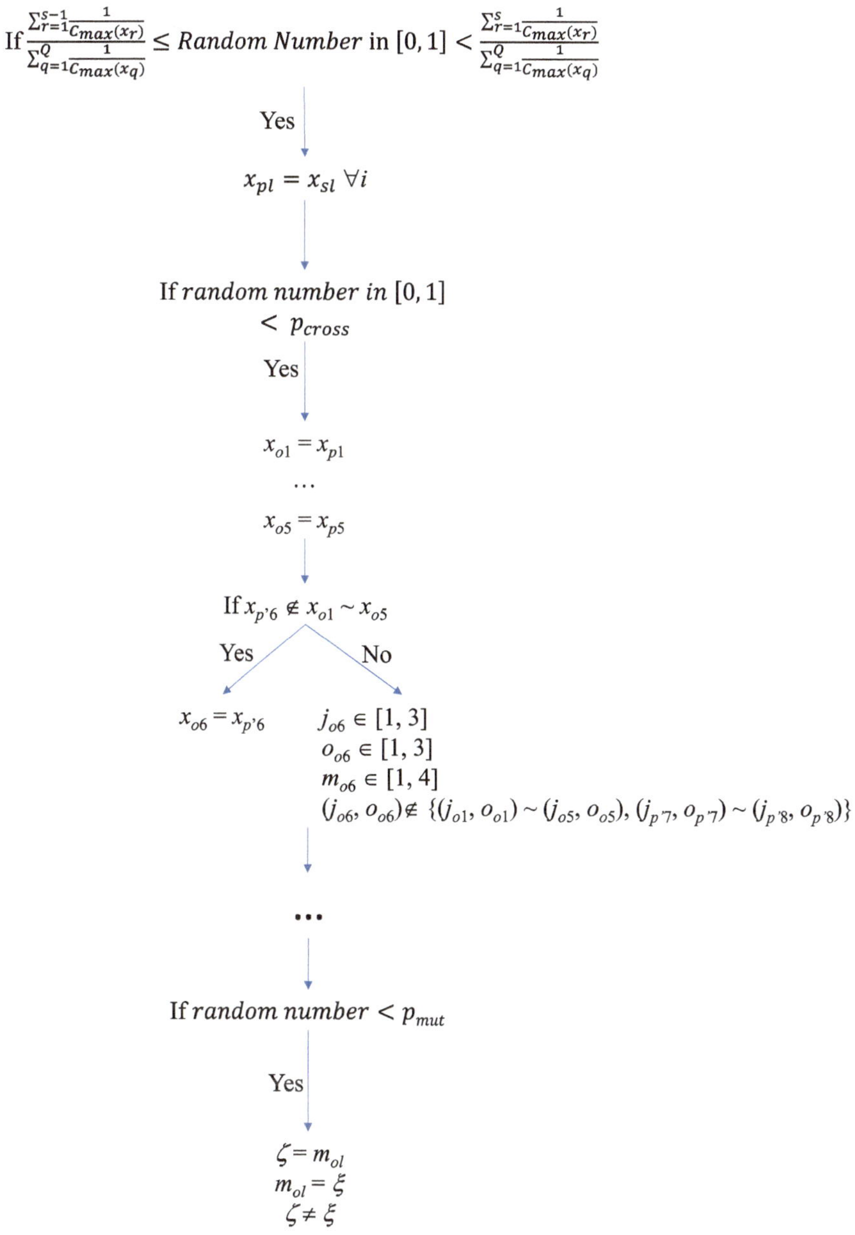

$$\text{If } \frac{\sum_{r=1}^{s-1} \frac{1}{C_{max}(x_r)}}{\sum_{q=1}^{Q} \frac{1}{C_{max}(x_q)}} \leq Random\ Number\ in\ [0,1] < \frac{\sum_{r=1}^{s} \frac{1}{C_{max}(x_r)}}{\sum_{q=1}^{Q} \frac{1}{C_{max}(x_q)}}$$

Yes

$$x_{pl} = x_{sl}\ \forall i$$

$$\text{If } random\ number\ in\ [0,1] < p_{cross}$$

Yes

$$x_{o1} = x_{p1}$$
$$\ldots$$
$$x_{o5} = x_{p5}$$

$$\text{If } x_{p'6} \notin x_{o1} \sim x_{o5}$$

Yes   No

$$x_{o6} = x_{p'6} \qquad j_{o6} \in [1,3]$$
$$o_{o6} \in [1,3]$$
$$m_{o6} \in [1,4]$$
$$(j_{o6}, o_{o6}) \notin \{(j_{o1}, o_{o1}) \sim (j_{o5}, o_{o5}), (j_{p'7}, o_{p'7}) \sim (j_{p'8}, o_{p'8})\}$$

$\bullet\bullet\bullet$

$$\text{If } random\ number < p_{mut}$$

Yes

$$\zeta = m_{ol}$$
$$m_{ol} = \xi$$
$$\zeta \neq \xi$$

Fig. 4.30 Decision tree for explaining the complete evolutional process of GA

Fig. 4.31 Radar chart for
visualizing the schedule

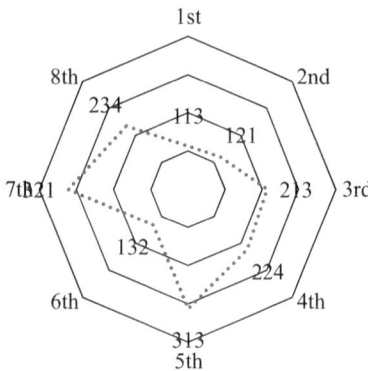

Subsequently, in order to visualize how the change of feasible solutions leads
to the optimal solution, the concept of scatter radar diagrams [19] is borrowed to
propose dynamic transition and contribution diagrams. In a dynamic transition and
contribution diagram,

- The data of all feasible solutions (i.e., chromosomes) of each population and the
  optimal solution are presented in the form of radar diagrams.
- The optimal solution is placed in the center.
- If a population contains numerous feasible solutions, only the feasible solutions
  that are the closest to the optimal solution will be liberated around the optimal
  solution.
- Unidirectional arrows link each feasible solution to the optimal solution.
- The thicker the arrow, the closer the feasible solution is to the optimal solution.
  Namely, the feasible solution contributes more to the optimal solution.

An example is given in Fig. 4.32 to show the dynamic transition and contribution
diagram at the end of the evolutional process.

4.3.6 SHAP Analysis

Extreme gradient boosting (XGBoost) can be applied to fit the relationship between
a feasible solution and the fitness. To this end, the last 1000 feasible solutions in
the GA, from the last several populations, have been recorded (see Fig. 4.33). The
Python code for implementing the XGBoost application is provided in Fig. 4.34.

Based on the fitting mechanism, SHAP analysis can be performed, as shown in
Figs. 4.35, 4.36, 4.37 and 4.38:

- M7 (the machine to process the 7th operation in the scheduling plan) is the most
  important inputs.
- The SHAP value of M7 is higher when its value increases.

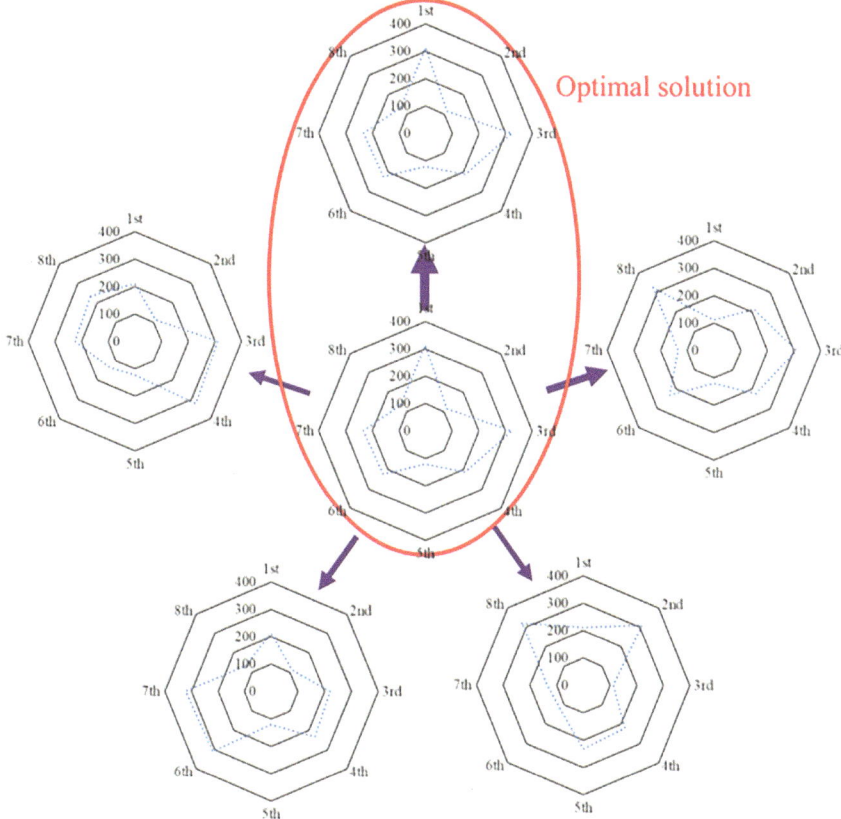

Fig. 4.32 A dynamic transition and contribution diagram

- O8 (the 8th operation) has the highest interaction with M7.
- In Fig. 4.37, all inputs have positive SHAP values.
- For this schedule, the most important inputs are M3 = 4 (the machine for processing the 3rd operation is machine #4) and M6 = 4 (the machine for processing the 6th operation is machine #4).

4.4 Evaluating and Comparing the Explanation Performances of Various XAI Methods

Various XAI techniques for explaining GA applications in job scheduling are compared in Table 4.3 in terms of their performances. Most existing XAI techniques can only meet up to five requirements. There is no perfect XAI technique to explain the scheduling process and result. Furthermore, existing XAI techniques

c	J1	O1	M1	J2	O2	M2	J3	O3	M3	J4	O4	M4	J5	O5	M5	J6	O6	M6	J7	O7	M7	J8	O8	M8	J9	O9	M9	fitness
1	3	1	4	3	2	4	3	3	3	1	1	4	1	2	2	2	1	1	1	3	2	2	2	2	2	3	2	0.02564
2	2	1	1	1	1	3	2	2	4	1	2	3	2	3	3	1	3	1	3	1	3	3	2	3	3	3	3	0.04545
3	1	1	1	1	2	3	3	1	1	2	1	2	1	3	2	2	2	4	2	3	2	3	2	3	3	3	3	0.04167
4	2	1	1	2	2	3	1	1	3	2	3	1	3	1	3	1	2	2	1	3	2	3	2	3	3	3	4	0.03704
5	1	1	3	1	2	3	2	1	3	3	1	4	3	2	4	3	3	1	2	2	4	2	3	3	1	3	1	0.05556
6	1	1	1	2	1	2	2	2	1	1	2	2	3	1	4	2	3	4	1	3	3	3	2	1	3	3	2	0.05263
7	1	1	1	1	2	2	3	1	3	1	3	1	2	1	4	2	2	3	2	3	3	3	2	2	3	3	3	0.04167
8	3	1	3	2	1	1	2	2	1	1	1	4	2	3	1	3	2	3	1	2	1	3	3	3	1	3	4	0.03846
9	2	1	1	1	1	1	2	2	1	2	3	2	3	1	4	1	2	1	3	2	4	3	3	3	1	3	3	0.04762
10	3	1	2	2	1	1	3	2	3	2	2	4	1	1	4	2	3	4	3	3	2	1	2	2	1	3	4	0.03448
11	3	1	3	3	2	4	3	3	2	2	1	4	2	2	3	1	1	2	1	2	2	1	3	1	2	3	2	0.03448
12	3	1	2	3	2	3	2	1	3	3	3	1	2	2	2	1	1	4	1	2	1	1	3	3	2	3	3	0.04167
13	2	1	2	2	2	2	1	1	1	3	1	4	1	2	1	1	3	2	3	2	2	2	3	4	3	3	3	0.04762
14	1	1	3	1	2	4	2	1	3	1	3	2	3	1	1	3	2	1	3	3	4	2	2	4	2	3	2	0.04545
15	2	1	2	1	1	3	2	2	3	1	2	1	2	3	2	1	3	3	3	1	1	3	2	3	3	3	4	0.04
16	3	1	2	2	1	4	3	2	2	2	2	4	3	3	1	2	3	4	1	1	3	1	2	3	1	3	1	0.05
17	1	1	2	2	1	4	3	1	3	1	2	4	2	2	3	3	2	2	2	3	3	3	3	1	1	3	2	0.05882
18	3	1	2	3	2	3	3	3	3	1	1	3	2	1	2	1	2	2	2	2	2	2	3	3	1	3	1	0.02857
19	3	1	4	2	1	1	1	1	4	2	2	2	1	2	1	3	2	3	2	3	1	1	3	1	3	3	3	0.03125

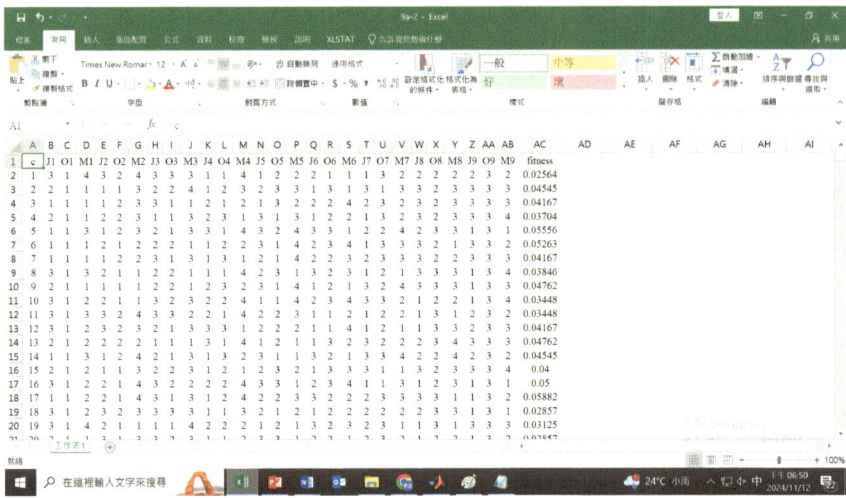

Fig. 4.33 Inputs and actual values of XGBoost

```
import xgboost
import shap
import pandas as pd
import numpy as np

allschedules = pd.read_excel('F:\\XAI-class\\講義\\9a-1.xlsx ')

X=allschedules.iloc[0:1000,1:28]
y=allschedules.iloc[0:1000,28]
y=y.to_numpy()
model1 = xgboost.XGBRegressor().fit(X, y)
O = model1.predict(X)

explainer1 = shap.Explainer(model1)
shap_values1 = explainer1(X)
# Compare the SHAP values of various inputs
shap.plots.beeswarm(shap_values1)
```

Fig. 4.34 Python code for implementing the XGBoost application

can be improved, and new (or tailored) ones can be proposed. Although each has its advantages, XAI techniques VII–IX can compensate for the existing methods' shortcomings and offer a comprehensive solution to explain GA operations.

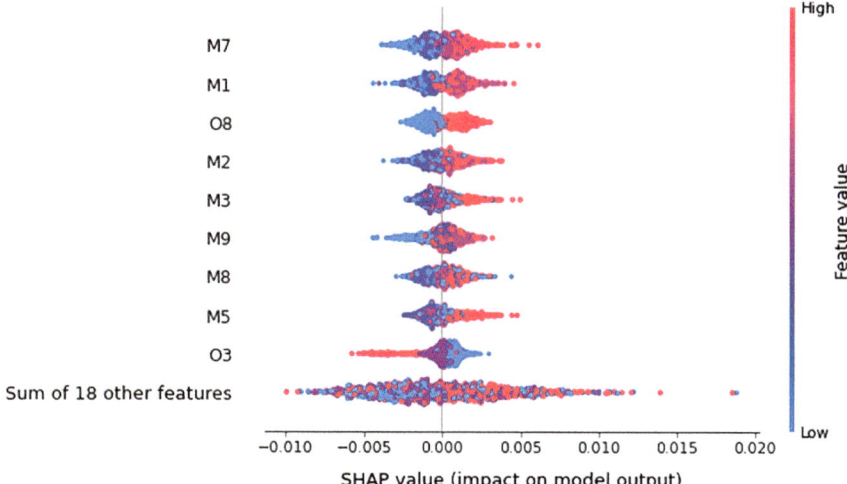

Fig. 4.35 Beeswarm plot for summarizing the results of the SHAP analysis for GA

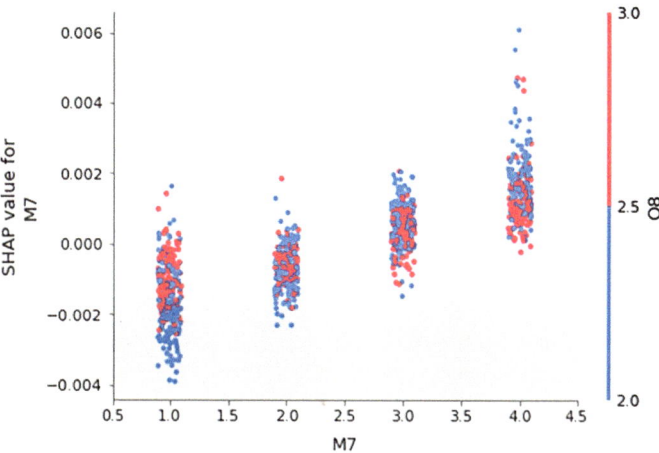

Fig. 4.36 Scatter plot for summarizing the results of the SHAP analysis for GA

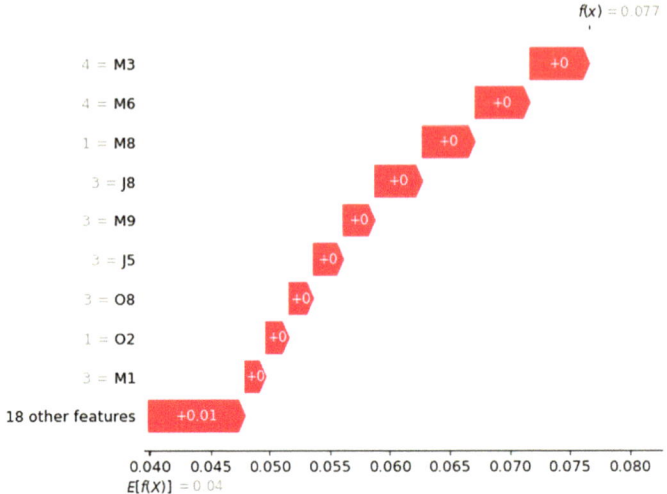

Fig. 4.37 Tornado plot for summarizing the results of a SHAP analysis (the optimal solution) for GA

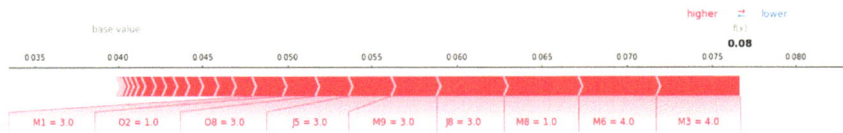

Fig. 4.38 Force plot for the same purpose

Table 4.3 Comparison of the explanation performances of various XAI techniques

	I	II	III	IV	V	VI	VII	VIII	IX	X
Requires background knowledge	V	V		V	V	V	V	V	V	V
Can handle high-dimensional data				V				V	V	
Compares scheduling performances						V	V			V
Has consistent explanation format				V	V	V	V	V	V	V
Is easy to communicate	V	V		V	V	V	V	V	V	V
Is easy to understand	V	V		V	V	V	V	V	V	V
Explains crossover operation	V	V	V	V				V		
Explains mutation operation	V	V	V	V				V		
Explains selection operation	V	V	V	V				V		
Visualizes contribution of inputs									V	V
Visualizes/compares feasible solutions				V					V	
Visualizes/traces evolution process				V	V				V	V

I: Textual description; II: Flowchart; III: Pseudocode; IV: Chromosome diagram; V: Dynamic line chart; VI: Bar chart; VII: Bar chart; VIII: Decision tree-based interpretation; IX: Dynamic transition and contribution diagram; X: Saliency map

References

1. B.O. Kong, M.S. Kim, B.H. Kim, J.H. Lee, Prediction of creep life using an explainable artificial intelligence technique and alloy design based on the genetic algorithm in creep-strength-enhanced ferritic 9% Cr steel. Metals Mater. Int. 1–12 (2022)
2. S.S. Sana, H. Ospina-Mateus, F.G. Arrieta, J.A. Chedid, Application of genetic algorithm to job scheduling under ergonomic constraints in manufacturing industry. J. Ambient. Intell. Humaniz. Comput. **10**(5), 2063–2090 (2019)
3. C. Shen, L. Wang, Q. Li, Optimization of injection molding process parameters using combination of artificial neural network and genetic algorithm method. J. Mater. Process. Technol. **183**(2–3), 412–418 (2007)
4. B. Skinner, S. Yuan, S. Huang, D. Liu, B. Cai, G. Dissanayake, H. Lau, A. Bott, D. Pagac, Optimisation for job scheduling at automated container terminals using genetic algorithm. Comput. Ind. Eng. **64**(1), 511–523 (2013)
5. J. Xia, Y. Yan, L. Ji, Research on control strategy and policy optimal scheduling based on an improved genetic algorithm. Neural Comput. Appl. **34**(12), 9485–9497 (2022)
6. B.M. Baker, M.A. Ayechew, A genetic algorithm for the vehicle routing problem. Comput. Oper. Res. **30**(5), 787–800 (2003)
7. Y.-C. Wang, T.-C.T. Chen, M.-C. Chiu, An explainable deep-learning approach for job cycle time prediction. Decision Analytics **6**, 100153 (2023)
8. T. Chen, C.-W. Lin, Smart and automation technologies for ensuring the long-term operation of a factory amid the COVID-19 pandemic: an evolving fuzzy assessment approach. Int. J. Adv. Manuf. Technol. **111**, 3545–3558 (2020)
9. L. Davis, Job shop scheduling with genetic algorithms, in *Proceedings of the first International Conference on Genetic Algorithms and their Applications* (2014), pp. 136–140
10. M.A. Salido, J. Escamilla, A. Giret, F. Barber, A genetic algorithm for energy-efficiency in job-shop scheduling. Int. J. Adv. Manuf. Technol. **85**, 1303–1314 (2016)
11. R. Qing-dao-er-ji, Y. Wang, A new hybrid genetic algorithm for job shop scheduling problem. Comput. Oper. Res. **39**(10), 2291–2299 (2012)
12. L. De Giovanni, F. Pezzella, An improved genetic algorithm for the distributed and flexible job-shop scheduling problem. Eur. J. Oper. Res. **200**(2), 395–408 (2010)
13. P. Garg, A comparison between memetic algorithm and genetic algorithm for the cryptanalysis of simplified data encryption standard algorithm. arXiv preprint arXiv:1004.0574 (2010)
14. F. Pezzella, G. Morganti, G. Ciaschetti, A genetic algorithm for the flexible job-shop scheduling problem. Comput. Oper. Res. **35**(10), 3202–3212 (2008)
15. C.R. Reeves, A genetic algorithm for flowshop sequencing. Comput. Oper. Res. **22**(1), 5–13 (1995)
16. A. Seker, S. Erol, R. Botsali, A neuro-fuzzy model for a new hybrid integrated process planning and scheduling system. Expert Syst. Appl. **40**(13), 5341–5351 (2013)
17. T.-C.T. Chen, Job sequencing and scheduling, in *Production Planning and Control in Semiconductor Manufacturing* (2023), pp. 77–100
18. C. Panigutti, A. Perotti, D. Pedreschi, Doctor XAI: An ontology-based approach to black-box sequential data classification explanations, in *Proceedings of the 2020 Conference on Fairness, Accountability, and Transparency* (2020), pp. 629–639
19. T. Chen, Y.C. Wang, A two-stage explainable artificial intelligence approach for classification-based job cycle time prediction. Int. J. Adv. Manuf. Technol. **123**(5), 2031–2042 (2022)
20. M.T. Ribeiro, S. Singh, C. Guestrin, "Why should i trust you?" Explaining the predictions of any classifier, in *Proceedings of the 22nd ACM SIGKDD International Conference on Knowledge Discovery and Data Mining* (2016), pp. 1135–1144
21. O. Abedinia, N. Amjady, H. Zareipour, A new feature selection technique for load and price forecast of electrical power systems. IEEE Trans. Power Syst. **32**(1), 62–74 (2016)
22. S. Zhang, C. Wang, A. Zomaya, Multi-level explanation of deep reinforcement learning-based scheduling. arXiv preprint arXiv:2209.09645 (2022)

23. J.H. Huh, J. Hwa, Y.S. Seo, Hierarchical system decomposition using genetic algorithm for future sustainable computing. Sustainability **12**(6), 2177 (2020)

24. B. Subhash, Explainable AI: saliency maps (2022). https://medium.com/@bijil.subhash/explainable-ai-saliency-maps-89098e230100

25. B. Zhou, A. Khosla, A. Lapedriza, A. Oliva, A. Torralba, Learning deep features for discriminative localization, in *Proceedings of the IEEE Conference on Computer Vision and Pattern Recognition* (2016), pp. 2921–2929

26. C.V. Goldman, M. Baltaxe, D. Chakraborty, J. Arinez, C.E. Diaz, Interpreting learning models in manufacturing processes: towards explainable AI methods to improve trust in classifier predictions. J. Ind. Inf. Integr. **33**, 100439 (2023)

27. O.I. Tukel, W.O. Rom, S.D. Eksioglu, An investigation of buffer sizing techniques in critical chain scheduling. Eur. J. Oper. Res. **172**(2), 401–416 (2006)

28. V.G. Costa, S. Salcedo-Sanz, C.E. Pedreira, Efficient evolution of decision trees via fully matrix-based fitness evaluation. Appl. Soft Comput. **150**, 111045 (2024)

29. J.S. Chou, K.E. Chen, Optimizing investment portfolios with a sequential ensemble of decision tree-based models and the FBI algorithm for effective financial indicator analysis. Appl. Soft Comput. 111550 (2024)

30. V. Rodriguez-Galiano, M. Sanchez-Castillo, M. Chica-Olmo, M.J.O.G.R. Chica-Rivas, Machine learning predictive models for mineral prospectivity: an evaluation of neural networks, random forest, regression trees and support vector machines. Ore Geol. Rev. **71**, 804–818 (2015)

31. Y.C. Wang, T. Chen, Adapted techniques of explainable artificial intelligence for explaining genetic algorithms on the example of job scheduling. Expert. Syst. Appl. **237**(A), 121369 (2024)

Chapter 5
Explaining Other Bio-inspired Algorithm Applications in Job Sequencing and Scheduling

5.1 ACO Applications in Job Sequencing and Scheduling

Job sequencing scheduling is a basic and important task for every manufacturing system. Regardless of the size of the manufacturing system, scheduling jobs currently in and to be released to the manufacturing system provides the basis for planning production, transportation, and other supporting activities [1]. Artificial intelligence (AI) technologies can build a job scheduling system or prepare the inputs required by the system [2, 3]. Some job scheduling problems are too complex to apply common dispatching rules. Mathematical programming (MP) models are formulated and optimized instead for such job scheduling problems. However, some of these MP models are intractable (NP hand). As a result, AI methods, such as bio-inspired algorithms, have been extensively applied to help optimize these MP models [4, 5]. Despite the effectiveness of these AI applications, they are not always easy to understand and communicate, hampering their credibility and acceptability. The concept of explainable artificial intelligence (XAI) was therefore proposed to tackle this problem [6]. However, the application of XAI in job scheduling is still limited so far [5].

AI methods that are considered black boxes and are widely used in job scheduling include artificial neural networks (ANN), genetic algorithms (GA), artificial bee colonies (ABC), ant colony optimization (ACO), particle swarm optimization (PSO), and others [7–11]. According to McNamara [12], if the operations in these AI methods can be described and made more transparent using ensemble methods, random forecasts, decision trees (DTs), Bayesian networks, and sparse linear models, then these AI methods will be more explainable, communicable and credible. To this end, this chapter is focused on how to enhance the explainability of ACO applications in job sequencing and scheduling.

ACO originated from the behavior of ant colonies in nature when searching for food and has been applied to many fields to assist in solving optimization problems. The execution process of ACO is shown in Fig. 5.1.

Fig. 5.1 Execution process
of ACO

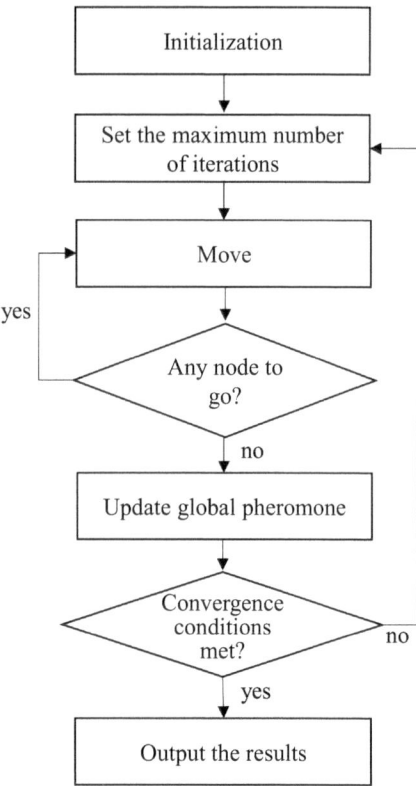

The key point of the algorithm is that when ants are looking for food, they leave pheromones in the paths they pass, thereby transmitting messages to other ants. Through pheromones, subsequent ants can find the shortest path to obtain food. The steps in Fig. 5.1 are explained in detail as follows.

First, the edge e_{ij} between each node and another node, its starting pheromone value τ_{ij} and the number of iterations are initialized. The initial positions of ants are also determined according to the scheduling problem.

Second, ants select and move to the next node in order to achieve the task. When located at node ξ (corresponding to operation o_ξ of job j_ξ on machine m_ξ), the probability of ant ρ moving to the next node ζ (corresponding to operation o_ζ of job j_ζ on machine m_ζ) is calculated based on the pheromone value ($\tau_{\xi\zeta}$):

$$p_{\xi\zeta}^{\rho} = \frac{\tau_{\xi\zeta}^{\alpha} \cdot \eta_{\xi\zeta}^{\beta}}{\sum_{l \in \mathbf{A}_{\xi}^{\rho}} (\tau_{\xi\zeta}^{\alpha} \cdot \eta_{\xi\zeta}^{\beta})} \tag{5.1}$$

where $\mathbf{A}_{\xi}^{\rho}$ is the set of all feasible adjacent nodes when ant ρ is located at node ξ. In addition,

$$\eta_{\xi\zeta} = C/d_{\xi\zeta} \tag{5.2}$$

in which $d_{\xi\zeta}$ is the distance from node ξ to node ζ; C is a positive constant; α and β are mutually influencing parameters: α is the pheromone trail, and β is heuristic information. If $\alpha = 0$, the ant will only choose based on the distance ($d_{\xi\zeta}$) when selecting the next node; when $\beta = 0$, the judgment will be based entirely on the concentration of pheromones. Finally, the roulette wheel method is used to select the next node.

Third, according to the movement paths of ants, the quality of the solution can be evaluated. For an ACO application in job scheduling, the quality of a feasible solution can be evaluated based on the scheduling performance, such as the makespan.

Fourth, when all ants in this iteration have completed their search, the pheromones ($\tau_{\xi\zeta}$) on all edges ($e_{\xi\zeta}$) are updated based on the best solution in the ant colony. The evaporation rate of pheromones (v) must be considered to elevate the probability of searching for better solutions:

$$\tau_{\xi\zeta} \leftarrow (1 - v)\tau_{\xi\zeta} \tag{5.3}$$

where $0 < v \leq 1$. In addition,

$$\tau_{\xi\zeta} \leftarrow \tau_{\xi\zeta} + \sum_{\rho=1}^{W} \Delta\tau_{\xi\zeta}^{\rho} \quad \forall(\xi, \zeta) \in L \tag{5.4}$$

in which $\Delta\tau_{\xi\zeta}^{\rho}$ is the pheromone deposited by ant ρ on the edge from node ξ to node ζ.

The above steps are repeated until the maximum number of iterations is met, or the originally set target value is reached. Then, the scheduling results are generated based on the best solution finally obtained.

Some selected cases and examples of such applications are reviewed as follows.

Example 5.1 The flexible job shop scheduling problem studied by Pezzella et al. [8] is taken as an example to illustrate the applicability of ACO, which is a scheduling problem with 3 jobs and 4 machines, as shown in Table 5.1. Pezzella et al. applied GA to help solve the flexible job shop scheduling problem. In contrast, ACO is applied in the following to search for the optimal solution to the flexible job shop scheduling problem.

The processing times of the operations of jobs in this flexible job shop scheduling problem refer to Table 5.1. It can be seen that each operation (O_{ij}) can be performed on all machines, but the processing times on different machines are not equal, which is more flexible than a traditional job shop scheduling problem. The goal of this scheduling problem is to minimize the total manufacturing time (makespan) that is expressed as $C_{\max}$ [13, 14].

ACO is applied to help solve this scheduling problem. The initialization settings are shown in Table 5.2. The setting of other parameters refers to Heinonen and

Table 5.1 Flexible job shop scheduling problem studied by Pezzella et al. [8]

	M_1	M_2	M_3	M_4
O_{11}	7	6	4	5
O_{12}	4	8	5	6
O_{13}	9	5	4	7
O_{21}	2	5	1	3
O_{22}	4	6	8	4
O_{23}	9	7	2	2
O_{31}	8	6	3	5
O_{32}	3	5	8	3

Pettersson [15] in which ACO was used to solve a job shop scheduling problem and Xing et al. [16] in which the knowledge-based ACO was applied to solve a flexible job shop scheduling problem. The setting results are presented in Table 5.3, where W represents the number of ants, which is set to 32 to be consistent with the number of feasible solutions to the scheduling problem.

All ants start from the nest. The initial values of pheromones on all paths are set to zero, that is, all searches start from the first generation of ants. When calculating pheromones $(\tau_{\xi\zeta})$ deposition, the pheromones concentration Q left by the ants passing by is set to 100:

$$\tau_{\xi\zeta} = Q/d_{\xi\zeta} \tag{5.5}$$

In addition, the number of operations in this flexible job shop scheduling problem is not large. Hoping to increase the heuristic information to a higher extent so that the ants can search for more paths, β is therefore set slightly larger than α. In addition, the pheromone evaporation rate is set to 0.8. In this way, in each iteration, too

Table 5.2 Initialization settings of ACO

Parameter	Value
Starting points of ants	Nest
Starting pheromone value $(\tau_{\xi\zeta})$ on each edge $(e_{\xi\zeta})$	0
Number of iterations	32

Table 5.3 Parameter setting results

Parameter	Value	Meaning
W	32	Number of ants
Q	100	Deposition pheromone
α	1	Preference pheromone
β	2	Nearest node
v	0.8	Pheromone evaporation rate

Table 5.4 Optimal solution obtained using ACO

Method	Scheduling results	Optimal objective function value
ACO	(2, 1, 3), (1, 1, 3), (3, 1, 3), (2, 2, 4), (2, 3, 4), (3, 2, 1), (1, 2, 1), (1, 3, 3)	13

Fig. 5.2 Gantt chart for illustrating the scheduling results

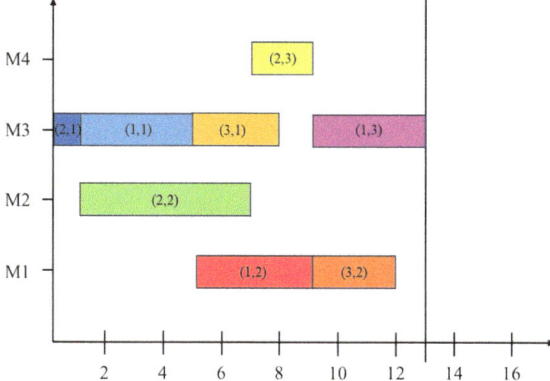

much pheromone is not left to the next iteration to prevent the ACO algorithm from prematurely converging to a single path.

The optimal solution obtained using ACO is shown in Table 5.4. For example, the first operation is the 1st operation of job #2 on machine #3. A Gantt chart is drawn to illustrate the scheduling results in Fig. 5.2, in which the information in brackets represents the job no. and operation no. Gantt charts are the most popular domain-specific tool for interpreting scheduling results [17, 18].

5.2 Selected Cases and Examples

The first motivation for explaining ACO applications in job sequencing and scheduling using XAI techniques and tools is that ACO reveals its effectiveness (or potential) in solving various job scheduling problems.

For example, Udhayakumar and Kumanan [19] attempted to minimize the makespan of a flexible manufacturing system (FMS) using ACO, where job operations can be performed on any of multiple machines with varying processing times. Li et al. [20] applied ACO to help solve an identical parallel batch processing machine (PBPM) scheduling problem, where multiple jobs (called batches) can be processed together on any of multiple identical machines, as is typical in wafer fabrication. Minimizing the average cycle time is a common goal in scheduling such systems. For a similar problem, Zhang et al. [21] considered the situation where various tools were shared between machines and each operation could be processed using different tools. They proposed a max–min ACO algorithm, in which the pheromone along a path were restricted to a range, for which the upper and lower boundaries were updated separately at each iteration. Wu et al. [22] applied ACO to solve a multi-objective flexible job shop scheduling problem. The objective functions were compromised by forming their weighted sum to be minimized. However, such a treatment may be problematic since the objective functions had different units and ranges.

5.3 Motivation for Using XAI to Explain ACO Applications in Job Sequencing and Scheduling

There are two main motivations for using XAI to explain ACO applications in job sequencing and scheduling:

- The interpretability of ACO applications in job scheduling has not been improved through the use of XAI techniques.
- Wang and Chen [5] used DTs, a popular XAI technique, to explain the application of GA in job scheduling. The explainability and communicability of the GA application have been enhanced. This successful experience prompted the present study to design a similar methodology.

For these reasons, decision tree-based XAI methods can be applied for interpreting ACO applications in job sequencing and scheduling. The novelty of decision tree-based XAI methods resides in the following:

- Attempts to use XAI applications to enhance the explainability of ACO are still rare.
- Most of the past research on XAI applications in manufacturing has been to fit the relationship between the inputs and output of machine learning (ML) or deep learning (DL) methods in a post hoc manner, which is not suitable for job scheduling problems because the output indicates the optimal schedule independent of the inputs (i.e., feasible schedules). In addition, it is difficult to distinguish the interpretation of ACO from that of other bio-inspired algorithms (such as GAs).
- Compared to other possible alternatives such as flowcharts, pseudocode, and universal modeling language (UML) for the same purpose, DTs are much simpler in format and logic and require minimal knowledge for user understanding and communication.

5.4 Explaining ACO Applications in Job Sequencing and Scheduling

5.4.1 Traditional Techniques and Tools

Text descriptions are still the most common technique for explaining ACO applications in scheduling. An example is given in Fig. 5.3.

Disjunction diagrams are traditionally used to illustrate scheduling problems [23]. $G_{dis} = (V, A, E)$ is a disjunction diagram, where V is the set of nodes, A is the set of conjunctive (directed) arcs, and E is the set of disjunctive (undirected) arcs. Given an instance of the scheduling problem, the disjunctive graph G_{dis} is built as follows. For each operation o, a node v_o is introduced. For each pair of operations o, o' with $o \leq_d o'$ (meaning o is a direct predecessor of o'), a conjunctive arc $a_{o,o'}$ is introduced. Taking the flexible job shop scheduling problem considered by Pezzella et al. [8] as an example where three jobs, each with 2–3 operations, need to be selectively processed

---

First, the edge e_{ij} between each node and another node, its starting pheromone value τ_{ij} and the number of iterations are initialized. The initial positions of ants are also determined according to the scheduling problem.

Second, ants select and move to the next node in order to achieve the task. When located at node ξ (corresponding to operation o_ξ of job j_ξ on machine m_ξ) the probability of ant ρ moving to the next node ζ (corresponding to operation o_ζ of job j_ζ on machine m_ζ) is calculated based on the pheromone value ($\tau_{\xi\zeta}$).

$$p_{\xi\zeta}^\rho = \frac{\tau_{\xi\zeta}{}^\alpha \cdot \eta_{\xi\zeta}{}^\beta}{\sum_{l \in A_\xi^\rho}(\tau_{\xi\zeta}{}^\alpha \cdot \eta_{\xi\zeta}{}^\beta)} \tag{2}$$

where A_ξ^ρ is the set of all feasible adjacent nodes when ant ρ is located at node ξ. In addition,

$$\eta_{\xi\zeta} = C/d_{\xi\zeta} \tag{3}$$

in which $d_{\xi\zeta}$ is the distance from node ξ to node ζ; C is a positive constant; α and β are mutually influencing parameters: α is the pheromone trail, and β is heuristic information. If $\alpha = 0$, the ant will only choose based on the distance ($d_{\xi\zeta}$) when selecting the next node; when $\beta = 0$, the judgment will be based entirely on the concentration of pheromones. Finally, the roulette wheel method is used to select the next node.

Fig. 5.3 Textual description for explaining a ACO application in scheduling

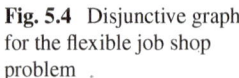

Fig. 5.4 Disjunctive graph for the flexible job shop problem .

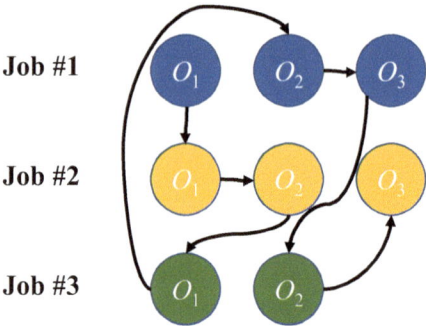

on four machines. The disjunction diagram for this scheduling problem is shown in Fig. 5.4.

Pseudocode is another tool for explaining the application of ACO to a scheduling problem [24], as shown in Fig. 5.5.

Gantt charts are undoubtedly the most popular tool for interpreting scheduling results, as shown in Fig. 5.2.

Similar to the application of GAs, dynamic line charts can also be drawn during the evolution/optimization process to observe whether the evolution process converges to the optimal solution [11] (see Fig. 5.6).

Comparing the performance of an AI application to those of other existing methods used for the same purpose is another way to interpret the AI application, as illustrated in the example below.

Example 5.2 Three heuristics are also applied to the scheduling problem for comparison: the exhaustive search method, the shortest processing time (SPT) rule, and the shortest remaining processing time (SRPT) rule. However, the due dates of jobs were not assigned in the scheduling problem, so dispatch rules such as earliest due date (EDD) and critical ratio (CR) do not apply [25, 26]. In the exhaustive search method, the sequences of all operations are randomly generated, but only eligible

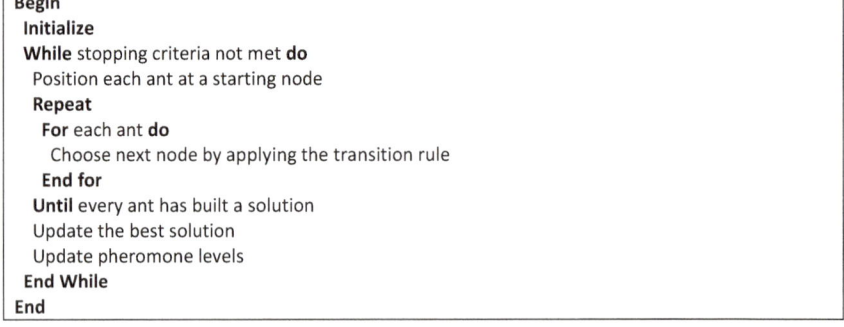

Fig. 5.5 Pseudocode for explaining the ACO application to a scheduling problem

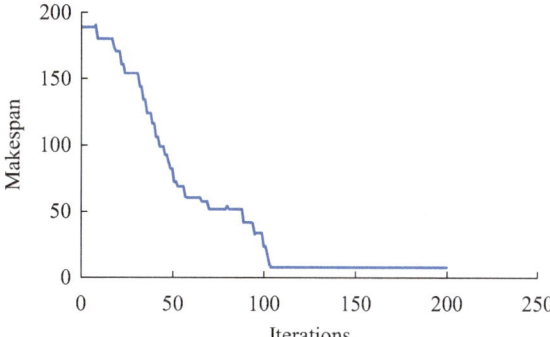

Fig. 5.6 Dynamic line chart for observing whether the evolution process converges to the optimal solution

sequences that adhere to the order of operations for each job are evaluated. The SPT scheduling rule selects the next operation with the shortest processing time as the next scheduling target. Therefore, the processing times of the next operations of jobs on the fastest machines are considered and compared. The SRPT scheduling rule first selects the job with the smallest sum of processing times for unprocessed operations and then performs the next operation of the job on the fastest machine [27]. The final (or best) solutions obtained by the three existing heuristics and their corresponding schedules are summarized in Table 5.5. Gantt charts for illustrating these schedules are presented in Fig. 5.7. In Fig. 5.7, when the exhaustive search method is applied, the optimal scheduling performance is consistent with that of ACO or GA [8], but the optimal scheduling results of these methods are different, showing that the scheduling problem has multiple optimal solutions.

When SPT is applied, the operation with the minimum processing time is always selected, so operations are concentrated to be performed on machines #3 and #4. In the case of centralized dispatching, the optimal scheduling performance only reaches 16. SRPT suffers from the same problem.

Table 5.5 Final/best solutions obtained using various methods

Method	Scheduling results	Final/best objective function value
Exhaustive search method	(2, 1, 3), (1, 1, 3), (3, 1, 4), (2, 2, 4), (1, 2, 1), (1, 3, 3), (3, 2, 2), (2, 3, 4)	13
SPT	(2, 1, 3), (3, 1, 3), (3, 2, 4), (1, 1, 3), (1, 2, 1), (1, 3, 3), (2, 2, 4), (2, 3, 4)	16
SRPT	(3, 1, 3), (3, 2, 1), (2, 1, 3), (2, 2, 4), (2, 3, 4), (1, 1, 3), (1, 2, 1), (1, 3, 3)	16

Fig. 5.7 Gantt charts for illustrating the scheduling results using the existing methods

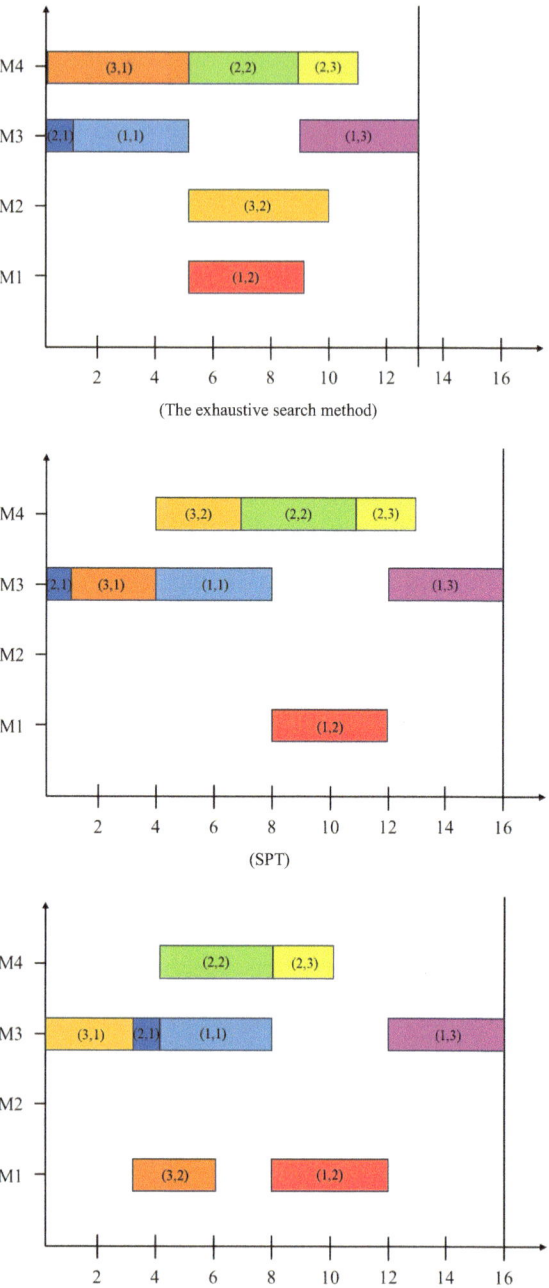

5.4.2 XAI Techniques and Tools

Attempts to explain ML or DL applications have been around for a long time. For example, confusion (or coincidence) matrices have been widely used to evaluate the classification performance using a ML (or DL) application. Tirkel [28] constructed a feedforward neural network (FNN) with two hidden layers to predict the cycle times of jobs and then divided them into several levels. Then, a confusion matrix was built to evaluate the classification performance, which reflected the accuracy of job cycle time prediction. Confusion (or coincidence) matrices can also be constructed to interpret the scheduling results using ACO. For example, before scheduling, jobs can be classified according to their due dates (or expected output times). The scheduling results using ACO may not conform to the expectations, for which a confusion matrix can be constructed, as shown in Fig. 5.8.

For a similar purpose, Chen and Wang [29] used fuzzy c-means (FCM) to classify jobs before using an ANN to predict their cycle times, which contained two tasks that require different interpretation methods to be applied. For the first task, a scatter radar chart was drawn to visualize the classification results for engineers to confirm whether the classification mechanism was appropriate. Wang and Chen [5] also visualized schedule differences using radar charts (see Fig. 5.9). For example, in Fig. 5.9, in the schedule on the left, operation (2, 1, 4) is executed first, while in that on the right, operation (2, 1, 1) is executed first.

Subsequently, for the second task, Chen and Wang constructed a random forest (RF) to use DTs to describe the behavior of the prediction mechanism (i.e., the ANN) in each local region. Senoner et al. [30] applied gradient boosting decision

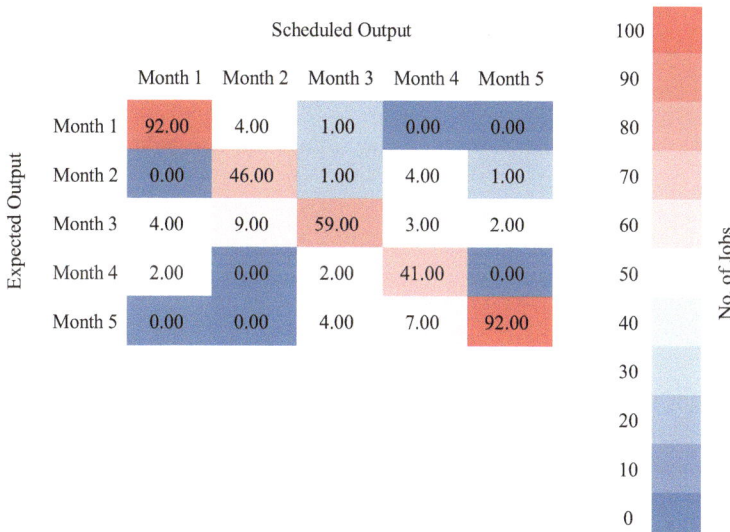

Fig. 5.8 Confusion matrix for interpreting the scheduling results using ACO

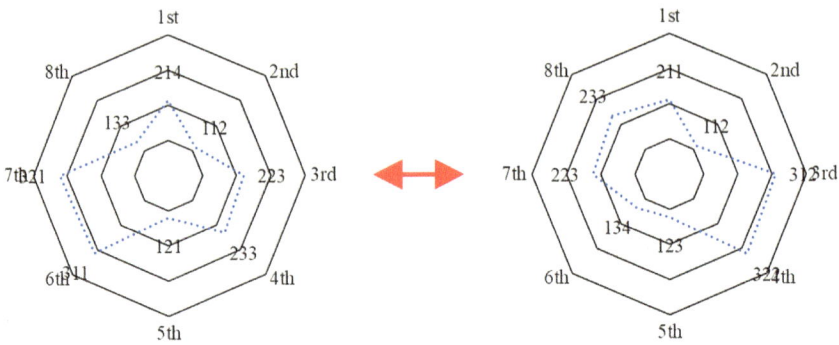

Fig. 5.9   Visualizing schedule differences with radar charts

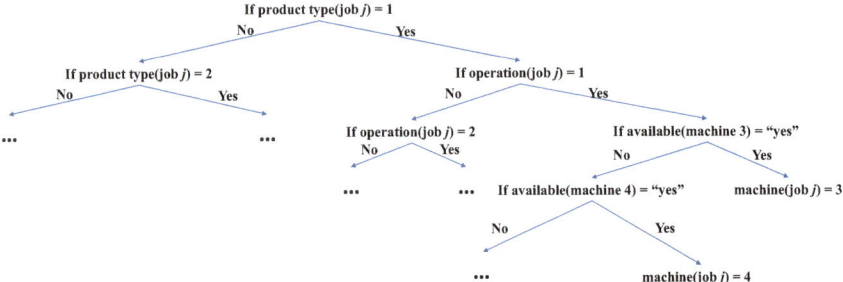

Fig. 5.10   DT for explaining the scheduling mechanism of ACO or other scheduling or optimization methods

tree (GBDT), also composed of DTs, and other ML methods to predict the yield of a transistor chip from many production parameter values, so as to improve the product quality in terms of the average and standard deviation of yield. Similar methods have been widely used to explain the application of DL in various fields, e.g., Wang et al. [31], Chen et al. [32], etc. A DT can also be constructed to explain the scheduling mechanism of ACO or other scheduling or optimization methods. An example is given in Fig. 5.10 for explaining the scheduling mechanism for the flexible job shop problem in Pezzella et al. [8].

5.4.3   SHAP Analysis

SHAP analysis is obviously the most popular XAI technique for distinguishing the impacts of inputs on an AI application. For example, in Kong et al. [33], an ANN was constructed to study the mechanism that the composition of an alloy affected its performance. In addition, Akhlaghi et al. [34] also constructed a deep neural network (DNN) to study how the features of a dew point cooler affected its performance.

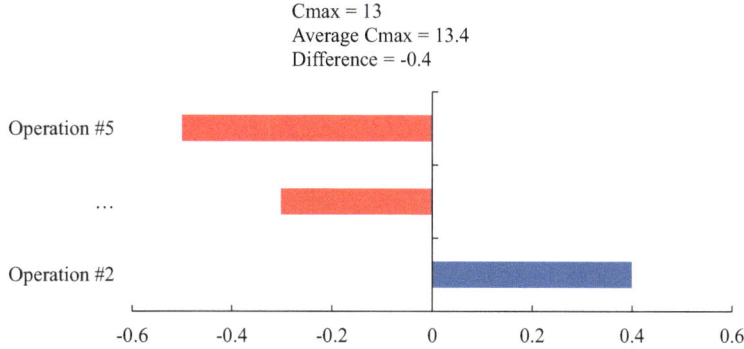

Fig. 5.11 SHAP analysis for distinguishing the impacts of operations in a schedule on the scheduling performance

These ANNs or DNNs were then applied to derive the SHAP value of each input to quantify its effect on the output. Similarly, SHAP analysis was performed by Senoner et al. [30] to determine the production parameters most critical to the transistor chip quality. SHAP analysis can indeed be applied to distinguish the impacts of operations in a schedule (obtained using ACO) on the scheduling performance, as illustrated in Fig. 5.11. In this figure, operation #5 has the greatest impact on the scheduling performance, as keeping it constant and randomizing the other operations reduces $C_{\max}$ by -0.5, which is a favorable property.

Extreme gradient boosting (XGBoost) can be applied to fit the relationship between a feasible solution and the makespan using ACO. Based on the fitting mechanism, SHAP analysis can be performed, as shown in Figs. 5.12, 5.13, 5.14 and 5.15.

5.4.4 Saliency Maps

Saliency maps [35] can be applied to associate points in the dynamic line chart with different colors to distinguish their performances in several ways. First, data points can be associated with different colors according to their performances:

$$\left[R_{p_t}, G_{p_t}, B_{p_t}\right] = \left[255 \cdot \left(1 - \frac{p_t - \min_s p_s}{\max_s p_s - \min_s p_s} \right), 255 \cdot \frac{p_t - \min_s p_s}{\max_s p_s - \min_s p_s}, 0 \right]$$

(5.6)

where p_t is the performance/fitness of population t. In Fig. 5.16, better scheduling performances are represented by green dots, while poor scheduling performances are represented by red dots. Thus, when a data point turns green, it indicates that the

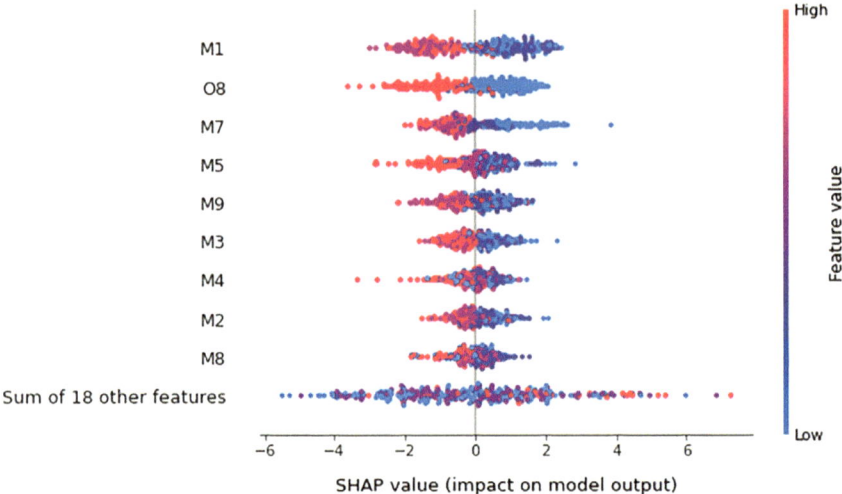

Fig. 5.12   Beeswarm plot for summarizing the results of the SHAP analysis for ACO

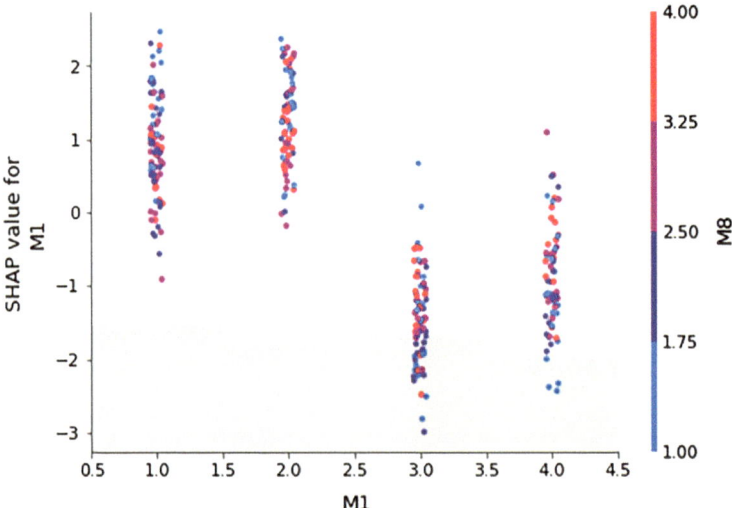

Fig. 5.13   Scatter plot for summarizing the results of the SHAP analysis for ACO

makespan has been reduced to a satisfactory level. Whether the evolution process needs to be terminated depends on how green the current data point is for the scheduler.

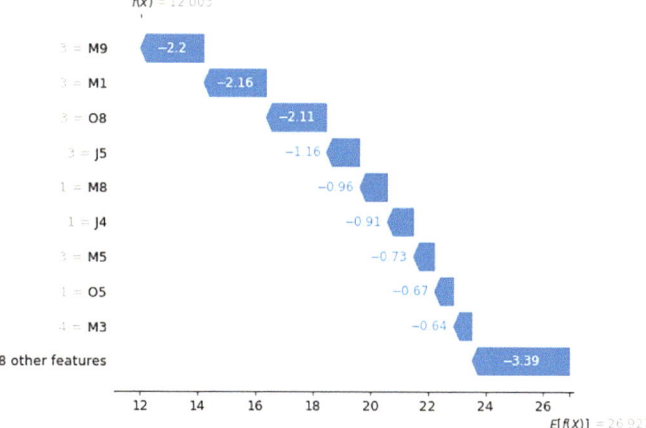

Fig. 5.14 Tornado plot for summarizing the results of a SHAP analysis (the optimal solution) for ACO

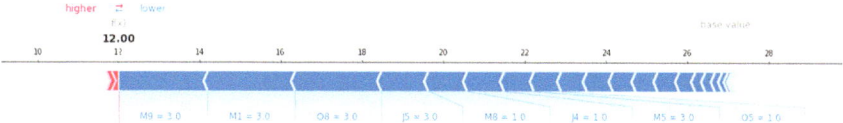

Fig. 5.15 Force plot for the same purpose

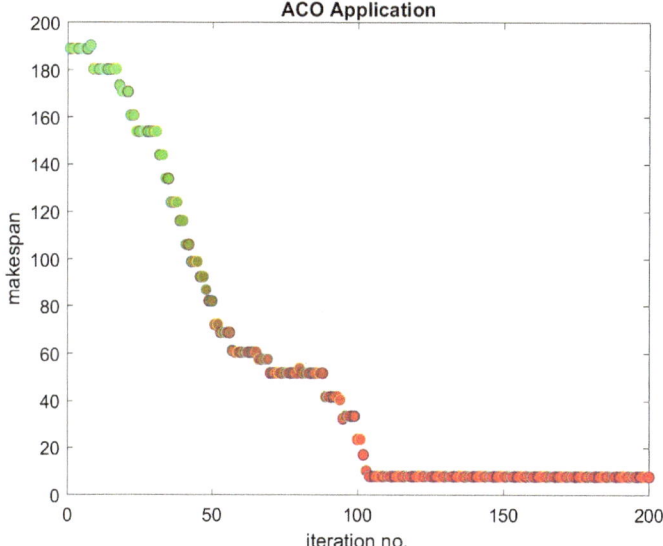

Fig. 5.16 Saliency map application to the dynamic line chart

The distinguishing effect can be further strengthened as follows:

$$R_{p_t} = 255 \cdot \left(\frac{\frac{p_{t+\Delta t} - p_t}{\Delta t} - \min\limits_{s}\left(\frac{p_{s+\Delta t} - p_s}{\Delta t}\right)}{\max\limits_{s}\left(\frac{p_{s+\Delta t} - p_s}{\Delta t}\right) - \min\limits_{s}\left(\frac{p_{s+\Delta t} - p_s}{\Delta t}\right)} \right)^{\varphi} \tag{5.7}$$

$$G_{p_t} = 255 \cdot \left(1 - \left(\frac{\frac{p_{t+\Delta t} - p_t}{\Delta t} - \min\limits_{s}\left(\frac{p_{s+\Delta t} - p_s}{\Delta t}\right)}{\max\limits_{s}\left(\frac{p_{s+\Delta t} - p_s}{\Delta t}\right) - \min\limits_{s}\left(\frac{p_{s+\Delta t} - p_s}{\Delta t}\right)} \right)^{\varphi} \right) \tag{5.8}$$

$$B_{p_t} = 0 \tag{5.9}$$

as illustrated in Fig. 5.17.

Saliency maps can also be applied to highlight the most important operations (using red bars) of the optimal schedule obtained using ACO. In contrast, greener bars indicate operations with less effects on the makespan. To this end, the gradient of the makespan/output with respect to the start time/input of each operation in the optimal solution can be calculated:

$$I_{o_l} = \frac{\partial C_{\max}}{\partial s_{o_l}} = \lim_{\Delta s_{o_l} \to 0} \frac{C_{\max}\left(s_{o_l} + \Delta s_{o_l}\right) - C_{\max}\left(s_{o_l}\right)}{\Delta s_{o_l}} \tag{5.10}$$

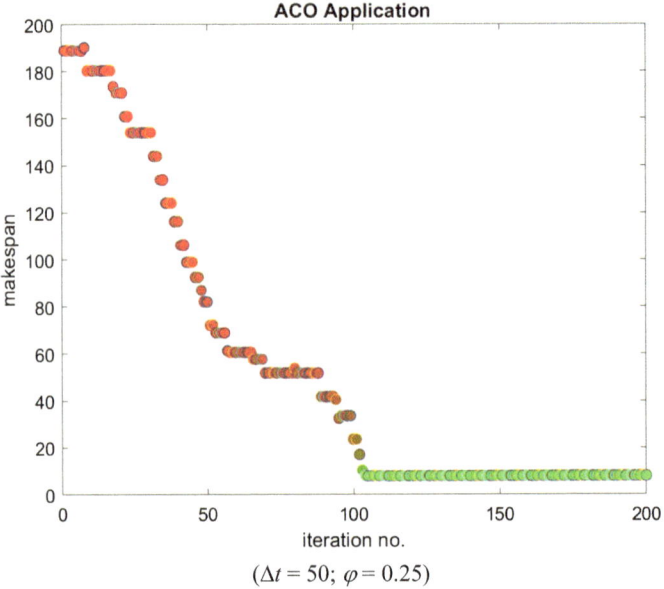

$$(\Delta t = 50; \ \varphi = 0.25)$$

Fig. 5.17 Results of strengthening the distinguishing effect

Then,

$$R_{p_l} = 255 \cdot \frac{\frac{C_{\max}(s_{o_l}+\Delta s_{o_l})-C_{\max}(s_{o_l})}{\Delta s_{o_l}} - \min\limits_{q}\left(\frac{C_{\max}(s_{o_q}+\Delta s_{o_q})-C_{\max}(s_{o_q})}{\Delta s_{o_q}}\right)}{\max\limits_{q}\left(\frac{C_{\max}(s_{o_q}+\Delta s_{o_q})-C_{\max}(s_{o_q})}{\Delta s_{o_q}}\right) - \min\limits_{q}\left(\frac{C_{\max}(s_{o_q}+\Delta s_{o_q})-C_{\max}(s_{o_q})}{\Delta s_{o_q}}\right)} \tag{5.11}$$

$$G_{p_l} = 255 \cdot \left(1 - \frac{\frac{C_{\max}(s_{o_l}+\Delta s_{o_l})-C_{\max}(s_{o_l})}{\Delta s_{o_l}} - \min\limits_{q}\left(\frac{C_{\max}(s_{o_q}+\Delta s_{o_q})-C_{\max}(s_{o_q})}{\Delta s_{o_q}}\right)}{\max\limits_{q}\left(\frac{C_{\max}(s_{o_q}+\Delta s_{o_q})-C_{\max}(s_{o_q})}{\Delta s_{o_q}}\right) - \min\limits_{q}\left(\frac{C_{\max}(s_{o_q}+\Delta s_{o_q})-C_{\max}(s_{o_q})}{\Delta s_{o_q}}\right)}\right) \tag{5.12}$$

$$B_{p_l} = 0 \tag{5.13}$$

The result is shown in Fig. 5.18, where $\Delta s_{o_l} = 1$; $l = 1 \sim 8$. Red operations are more important than green operations. Obviously, the operations in the critical chain have been highlighted.

To further distinguish the effects of operations, their importance levels are evaluated with the changes that should be made to their start times to worsen the scheduling performance. For this purpose, Eq. (5.13) is changed to

$$I_{o_l} = \frac{\Delta C_{\max}}{\Delta s_{o_l}} = \frac{C_{\max}(s_{o_l}+\Delta s_{o_l})-C_{\max}(s_{o_l})}{\Delta s_{o_l}} = \frac{C_{\max}(s_{o_l})+1-C_{\max}(s_{o_l})}{\Delta s_{o_l}} = \frac{1}{\Delta s_{o_l}} \tag{5.14}$$

The evaluation results are summarized in Table 5.6.

The color associated with an operation is determined as follows (see Table 5.7):

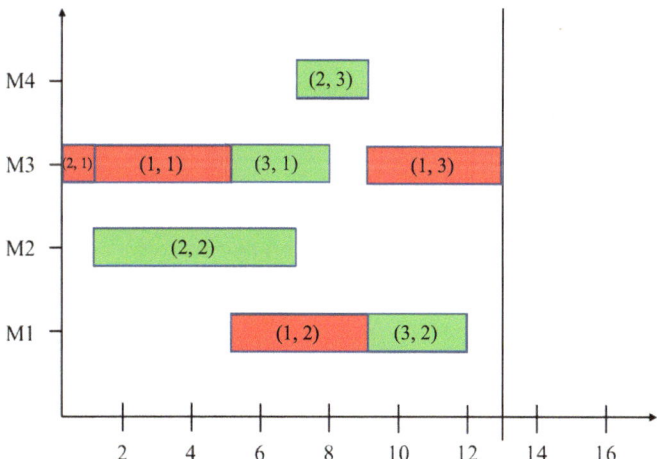

Fig. 5.18 Saliency map application to the optimal schedule obtained using ACO

Table 5.6 Operation importance evaluation results

l	j (Job)	o (Operation)	m (Machine)	I_{o_l}
1	2	3	4	1/5
2	2	1	3	1
3	1	1	3	1
4	3	1	3	1/2
5	1	3	3	1
6	2	2	2	1/5
7	1	2	4	1
8	3	2	4	1/2

Table 5.7 Colors of operations

l	j (Job)	o (Operation)	m (Machine)	R	G	B
1	1	2	1	255	0	0
2	3	2	1	0	255	0
3	1	3	2	137	118	0
4	1	1	3	255	0	0
5	2	1	3	118	137	0
6	3	1	3	0	255	0
7	2	2	4	59	196	0
8	2	3	4	0	255	0

$$R_{p_t} = 255 \cdot \left(\frac{I_{o_l} - \min\limits_{q}\left(I_{o_q}\right)}{\max\limits_{q}\left(I_{o_q}\right) - \min\limits_{q}\left(I_{o_q}\right)} \right)^{\varphi} \tag{5.15}$$

$$G_{p_t} = 255 \cdot \left(1 - \left(\frac{I_{o_l} - \min\limits_{q}\left(I_{o_q}\right)}{\max\limits_{q}\left(I_{o_q}\right) - \min\limits_{q}\left(I_{o_q}\right)} \right)^{\varphi} \right) \tag{5.16}$$

$$B_{p_t} = 0 \tag{5.17}$$

The salient map application results are shown in Fig. 5.19.

5.4.5 Decision Tree-Based XAI Approach

Among XAI methods, interpretation methods based on decision trees are one of the most commonly used interpretation methods. For example, Sagi and Rokac [36] proposed that converting GBDT into interpretable decision trees did not affect its

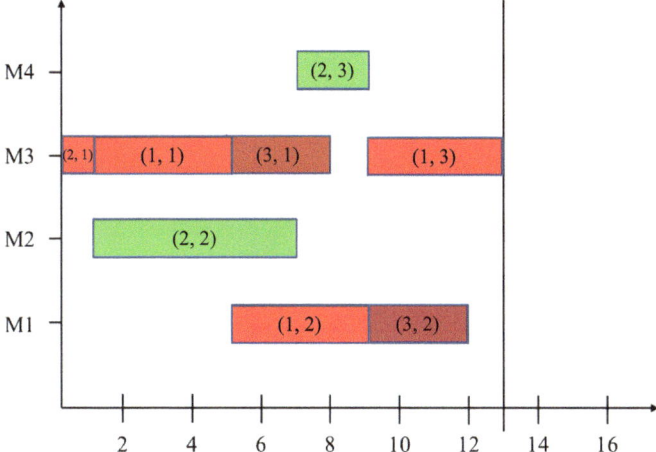

Fig. 5.19 Results of distinguishing the effects of operations

prediction accuracy. The results were also easier to understand for non-algorithm experts. As the most important application of AI in network security, intrusion detection systems (IDSs) are becoming more and more popular, but most of them are very complex and are widely regarded as black boxes. Mahbooba et al. [37] constructed decision trees to explain the inference mechanism of IDSs, so that malicious or regular network traffic could be distinguished by observing the characteristics of network traffic.

The transparencies of the ACO steps are compared. Based on the comparison results, the most opaque steps, the search for possible subsequent nodes (steps), and the selection from these possible subsequent nodes (steps) are to be interpreted with DTs.

For interpreting the step of screening all possible next nodes, a DT is constructed in Fig. 5.20, where j, o, and m represent job no., operation no., and machine no., respectively. Therefore, nodes are represented by (j, o, m), where $j \in [1, N]$, $o \in [1, S]$, and $m \in [1, M]$. $CONT$ is a state variable. If $CONT = 0$, the screening process is terminated; otherwise, the screening process continues. In addition, two state variables are defined as SE_{jom} and $\mathbf{A}$. If $SE_{jom} = 1$, node (j, o, m) has been selected; otherwise ($SE_{jom} = 0$), node (j, o, m) has not been selected. $\mathbf{A}$ is the set of all selected nodes. Before selecting the next node, $\mathbf{A}$ is an empty set, i.e., $\mathbf{A} = \phi$.

After explaining how an ant chooses the next possible node, how the ant chooses the next node it goes to needs to be explained. In this study, a roulette wheel selection method is applied to make a random but weighted selection among all selectable nodes. The roulette wheel selection method is also used in other bionic algorithms, such as GAs, to select parental chromosomes to be paired based on their fitness values, making it easier for better solutions to be selected without losing the exploratory nature of the algorithm. However, the weighting part in ACO is according to the probability calculated by p, for which the specific DT model is shown in Fig. 5.21,

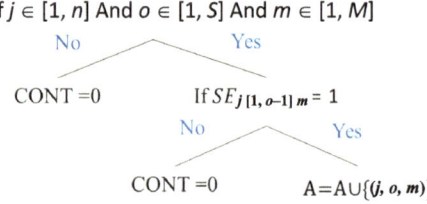

If $j \in [1, n]$ And $o \in [1, S]$ And $m \in [1, M]$

Fig. 5.20 Decision tree for interpreting the operation of screening all possible next nodes

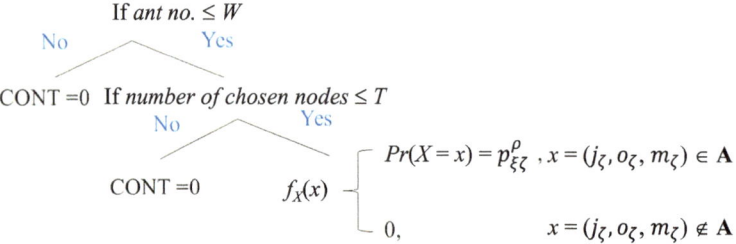

Fig. 5.21 DT model for explaining how an ant chooses the next node

where W represents the number of ants in each iteration, and T represents the number of nodes that can be selected in a trip. In addition, X indicates the selected node. According to the logic of the roulette wheel method, X is a probability mass function, in which the x-axis includes all feasible nodes in **A**, and the y-axis represents the p value of each feasible node.

Example 5.3 The decision tree-based XAI approach is applied to the case in Example 5.1 to explain the reasoning process of ACO as follows. First, N (the number of jobs) $= 3$; $S = 3$ (the maximum number of operations); $M = 4$ (the number of machines). To make the input data to the scheduling problem a complete matrix, job #3 is defined to have a (i.e., the third) virtual operation, and its processing times on all four machines are 0's. **A** is used to store all nodes that meet the conditions; SE_{jom} records the selected nodes; $j = 1 \sim 3$; $o = 1 \sim 3$; $m = 1 \sim 4$. The state variable $CONT$ shows whether the screening process continued or not. For the step of screening all possible next nodes, the interpretation results are shown in Fig. 5.22.

After screening all possible next nodes, the decision tree-based XAI approach is applied to explain how ants made choices, as shown in Fig. 5.23. In this figure, w, the number of ants in each iteration, should be less than (or equal to) the total number of ants $W = 32$; t, the number of nodes that each ant needs to pass when completing a search path should be less than the total number of operations $T = 9$. When an ant makes a choice, it first needs to ensure that the number of ants that have made their choices is within the set total number. Then, it also needs to confirm that the ant has not made a complete search. Finally, the feasible solutions stored in **A** are chosen using the roulette wheel method.

If $o = 1$ And $j \in [1, 3]$ And $m \in [1, 4]$

Yes / No

$\mathbf{A} = \mathbf{A} \cup \{j, 1, m\}$ If $o = 2$ And $j \in [1, 3]$ And $m \in [1, 4]$

Yes / No

If $SE_{j1m} = 1$ If $o = 3$ And $j \in [1, 3]$ And $m \in [1, 4]$

Yes / No   Yes

$\mathbf{A} = \mathbf{A} \cup \{j, 2, m\}$ $CONT = 0$ If $SE_{j1m} = 1$ And $SE_{j2m} = 1$

Yes / No

$\mathbf{A} = \mathbf{A} \cup \{j, 3, m\}$ $CONT = 0$

Fig. 5.22 Interpretation results for the step of screening all possible next nodes

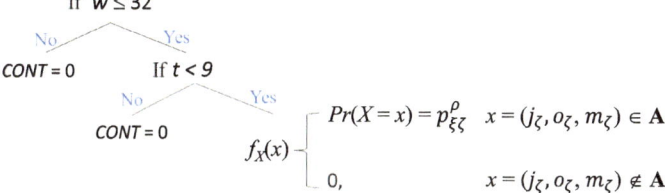

If $w \le 32$

No / Yes

$CONT = 0$ If $t < 9$

No / Yes

$CONT = 0$

$$f_X(x) = \begin{cases} Pr(X = x) = p_{\xi\zeta}^{\rho} & x = (j_{\zeta}, o_{\zeta}, m_{\zeta}) \in \mathbf{A} \\ 0, & x = (j_{\zeta}, o_{\zeta}, m_{\zeta}) \notin \mathbf{A} \end{cases}$$

Fig. 5.23 Application of the decision tree-based XAI approach for explaining how ants made choices

5.4.6 Post Hoc XAI Analysis

Based on the optimal scheduling results obtained using ACO, a post hoc XAI analysis can be conducted to evaluate the importance of an input on the output. Take the processing time of each operation as an example. Partial derivation (PR) can be applied to change the processing time of each operation by 1 in the optimal solution to observe the change of the optimal scheduling performance. The results are summarized in Table 5.8.

Table 5.8 Post hoc XAI analysis

Operation	Machine	C_{max} (Original)	C_{max} (Add 1 to processing time)	C_{max} (Subtract 1 from processing time)	Mean absolute change
O_{11}	#3	12	12	12	0
O_{12}	#1	12	13	11	1
O_{13}	#3	12	13	11	1
...					

Table 5.9 Comparison of three possible XAI methods for explaining the reasoning process of ACO

Method	Explanation consistency	Required background knowledge	Easiness to comprehend
Flowchart	Low	Little	High
UML	High	A lot	Low
Decision tree-based XAI approach	High	Little	High

5.5 Evaluating and Comparing the Explanation Performances of Various XAI Methods

The scheduling results using ACO can be well illustrated with a Gantt chart. In addition, three possible methods for explaining the reasoning process of ACO are compared in Table 5.9. Obviously, the decision tree-based XAI approach has high explanation consistency, is easy to comprehend, and requires little background knowledge.

In addition, ACO, GA, and the exhaustive search method all optimize the scheduling performance in terms of the minimum makespan, but the optimal scheduling results using these methods are different, indicating that there are multiple optimal solutions to the scheduling problem. Therefore, even if the same XAI tool is used, the interpretation results of these AI applications are not the same.

References

1. Y.C. Wang, T. Chen, C.W. Lin, A slack-diversifying nonlinear fluctuation smoothing rule for job dispatching in a wafer fabrication factory. Robot. Comput. Integr. Manuf. **29**(3), 41–47 (2013)
2. H. Wang, Flexible flow shop scheduling: optimum, heuristics and artificial intelligence solutions. Expert. Syst. **22**(2), 78–85 (2005)
3. T. Chen, Y.-C. Wang, A fuzzy-neural approach for supporting three-objective job scheduling in a wafer fabrication factory. Neural Comput. Appl. **23**(1), 353–367 (2013)
4. M. Del Gallo, G. Mazzuto, F.E. Ciarapica, M. Bevilacqua, Artificial intelligence to solve production scheduling problems in real industrial settings: systematic literature review. Electronics **12**(23), 4732 (2023)
5. Y.C. Wang, T. Chen, Adapted techniques of explainable artificial intelligence for explaining genetic algorithms on the example of job scheduling. Expert. Syst. Appl. **237**(A), 121369 (2024)
6. V. Bento, M. Kohler, P. Diaz, L. Mendoza, M.A. Pacheco, Improving deep learning performance by using explainable artificial intelligence (XAI) approaches. Discov. Artif. Intell. **1**, 1–11 (2021)
7. C.J. Liao, C.T. Tseng, P. Luarn, A discrete version of particle swarm optimization for flowshop scheduling problems. Comput. Oper. Res. **34**(10), 3099–3111 (2007)
8. F. Pezzella, G. Morganti, G. Ciaschetti, A genetic algorithm for the flexible job-shop scheduling problem. Comput. Oper. Res. **35**(10), 3202–3212 (2008)

9. R. Akbari, V. Zeighami, K. Ziarati, Artificial bee colony for resource constrained project scheduling problem. Int. J. Ind. Eng. Comput. **2**(1), 45–60 (2011)
10. W. Deng, J. Xu, H. Zhao, An improved ant colony optimization algorithm based on hybrid strategies for scheduling problem. IEEE Access **7**, 20281–20292 (2019)
11. T.-C.T. Chen, Job sequencing and scheduling, in *Production Planning and Control in Semiconductor Manufacturing* (2023), pp. 77–100
12. M. McNamara, Explainable AI: What is it? How does it work? And what role does data play? (2022). https://www.netapp.com/blog/explainable-ai/
13. T. Chen, A tailored non-linear fluctuation smoothing rule for semiconductor manufacturing factory scheduling. Proc. Inst. Mech. Eng. Part I J. Syst. Control. Eng. **223**(2), 149–160 (2009)
14. M.M. Ahmadian, M. Khatami, A. Salehipour, T.C.E. Cheng, Four decades of research on the open-shop scheduling problem to minimize the makespan. Eur. J. Oper. Res. **295**(2), 399–426 (2021)
15. J. Heinonen, F. Pettersson, Hybrid ant colony optimization and visibility studies applied to a job-shop scheduling problem. Appl. Math. Comput. **187**, 989–998 (2007)
16. L.N. Xing, Y.W. Chen, P. Wang, Q.S. Zhao, J. Xiong, A knowledge-based ant colony optimization for flexible job shop scheduling problems. Appl. Soft Comput. **10**(3), 888–896 (2010)
17. T. Chen, Y.C. Wang, A nonlinear scheduling rule incorporating fuzzy-neural remaining cycle time estimator for scheduling a semiconductor manufacturing factory—a simulation study. Int. J. Adv. Manuf. Technol. **45**, 110–121 (2009)
18. M. Shibuya, X. Chen, Production planning and management using Gantt charts. J. Mech. Eng. Autom. **11**, 68–76 (2021)
19. P. Udhayakumar, S. Kumanan, Sequencing and scheduling of job and tool in a flexible manufacturing system using ant colony optimization algorithm. Int. J. Adv. Manuf. Technol. **50**, 1075–1084 (2010)
20. L. Li, P. Gu, F. Qiao, Y. Wu, Q. Wu, ACO-based multi-objective scheduling of identical parallel batch processing machines in semiconductor manufacturing, in *Future Manufacturing Systems* (2010), pp. 163–178
21. X. Zhang, S. Wang, L. Yi, H. Xue, S. Yang, X. Xiong, An integrated ant colony optimization algorithm to solve job allocating and tool scheduling problem. Proc. Inst. Mech. Eng. Part B J. Eng. Manuf. **232**(1), 172–182 (2018)
22. J. Wu, G.D. Wu, J.J. Wang, Flexible job-shop scheduling problem based on hybrid ACO algorithm. Int. J. Simul. Model. **16**(3), 497–505 (2017)
23. C. Blum, M. Sampels, An ant colony optimization algorithm for shop scheduling problems. J. Math. Model. Algorithms **3**, 285–308 (2004)
24. D.N. Kumar, M.J. Reddy, Ant colony optimization for multi-purpose reservoir operation. Water Resour. Manage **20**, 879–898 (2006)
25. T. Chen, A. Jeang, Y.C. Wang, A hybrid neural network and selective allowance approach for internal due date assignment in a wafer fabrication plant. Int. J. Adv. Manuf. Technol. **36**, 570–581 (2008)
26. G.A. Rolim, M.S. Nagano, Structural properties and algorithms for earliness and tardiness scheduling against common due dates and windows: a review. Comput. Ind. Eng. **149**, 106803 (2020)
27. A. Al-Refaie, T. Chen, M. Judeh, Optimal operating room scheduling for normal and unexpected events in a smart hospital. Oper. Res. Int. Journal **18**, 579–602 (2018)
28. I. Tirkel, Cycle time prediction in wafer fabrication line by applying data mining methods, in *2011 IEEE/SEMI Advanced Semiconductor Manufacturing Conference* (2011), pp. 1–5
29. T. Chen, Y.C. Wang, A two-stage explainable artificial intelligence approach for classification-based job cycle time prediction. Int. J. Adv. Manuf. Technol. **123**(5), 2031–2042 (2022)
30. J. Senoner, T. Netland, S. Feuerriegel, Using explainable artificial intelligence to improve process quality: evidence from semiconductor manufacturing. Manage. Sci. **68**(8), 5704–5723 (2022)

31. Y.-C. Wang, T.-C.T. Chen, M.-C. Chiu, An explainable deep-learning approach for job cycle time prediction. Decis. Anal. **6**, 100153 (2023)
32. T.C.T. Chen, C.W. Lin, Y.C. Lin, A fuzzy collaborative forecasting approach based on XAI applications for cycle time range estimation. Appl. Soft Comput. **151**, 111122 (2024)
33. B.O. Kong, M.S. Kim, B.H. Kim, J.H. Lee, Prediction of creep life using an explainable artificial intelligence technique and alloy design based on the genetic algorithm in creep-strength-enhanced ferritic 9% Cr steel. Metals Mater. Int. 1–12 (2022)
34. Y.G. Akhlaghi, K. Aslansefat, X. Zhao, S. Sadati, A. Badiei, X. Xiao, S. Shittu, Y. Fan, X. Ma, Hourly performance forecast of a dew point cooler using explainable Artificial Intelligence and evolutionary optimisations by 2050. Appl. Energy **281**, 116062 (2021)
35. B. Subhash, Explainable AI: Saliency maps (2022). https://medium.com/@bijil.subhash/explainable-ai-saliency-maps-89098e230100
36. O. Sagi, L. Rokach, Approximating XGBoost with an interpretable decision tree. Inf. Sci. **572**, 522–542 (2021)
37. B. Mahbooba, M. Timilsina, R. Sahal, M. Serrano, Explainable artificial intelligence (XAI) to enhance trust management in intrusion detection systems using decision tree model. Complexity **2021**, 6634811 (2021)